PSYCHOLOGICAL CONSEQUENCES OF CHILD ABUSE

About the Author

Dr. Sushma Pandey is Reader in Psychology at D.D.U. Gorakhpur University, Gorakhpur. She received her Ph.D. in Psychology in 1987. Earlier she has served in the Department of Population Studies, S.V. University, Tirupati. Her areas of interest include Social Psychology, Community Psychology, Cognitive Processes and Child Abuse.

Dr. Pandey has worked extensively in the areas of child abuse, domestic violence, time perspective and cognitive processes, women empowerment, reproductive health, stress and coping. She has published numerous research articles in the national and international journals of repute.

Dr. Pandey has recently completed one mojor research project, entitled, *Social Dynamics of Child Abuse and its Psychological Consequences in Eastern Districts of U.P.* Also, she has completed six research projects under D.S.A. Programme. She has developed numerous measures to assess child abuse, domestic violence and women empowerment in Indian context. Currently, she is working on preventive strategies of child abuse/domestic violence and their consequences.

PSYCHOLOGICAL CONSEQUENCES OF CHILD ABUSE

Dr. (Smt.) Sushma Pandey
Department of Psychology (D.S.A)
D.D.U. Gorakhpur University
Gorakhpur-273009

CONCEPT PUBLISHING COMPANY, NEW DELHI-110059

Cataloging in Publication Data—DK
Courtesy: D.K. Agencies (P) Ltd. <docinfo@dkagencies.com>

Pandey, Sushma.
Psychological consequences of child abuse / Sushma Pandey.
p. cm.
Study conducted at Gorakhpur Division of Uttar Pradesh, India.
Includes bibliographical references (p.)
Includes index.
ISBN 8180694054

1. Child abuse—India—Gorakhpur Division —Psychological aspects.
I. Title.

DDC 362.760 954 2 22

ISBN : 81-8069-405-4 (HB)

First Published 2007

Published and Printed by
Ashok Kumar Mittal
Concept Publishing Company
A/15-16, Commercial Block, Mohan Garden
New Delhi-110059 (India)

Phones : 25351460, 25351794
Fax : 091-11-25357103
Email : publishing@conceptpub.com

FOREWORD

Child abuse as a socio-psychological construct in social sciences is fraught with ubiquity. Numerous social scientists including psychologists have investigated its varied forms, dynamics, and psychological consequences. Some of them have endeavoured to develop preventive programmes. However, the empirical referents of child abuse are still not systematized. As a result, the term 'child abuse' has been neither comprehensively characterized nor some suitable theory has been developed. In spite of the fact that innumerable number of empirical research articles have been published, it is clearly evident that each investigator has selected a few salient forms of child abuse, some fractional facets of background factors, and socio-psychological consequences in accordance with his/her tacit assumptions, methodological convenience, and temporal constraints on completion of the research project. Of course, there is no denying the fact that these researches taken together did provide a sizeable number of relevant variables about the background of abused children, child abusing parents, their environmental features, socio-economic strata and social ethos. It is interesting to note that most of the child abuse investigating social psychologists, wittingly or unwittingly omitted to investigate detailed psychological make up of abused children and abusing parents/guardians.

In this backdrop, the monograph entitled "Psychological Consequence of Child Abuse", and authored by Sushma Pandey presents an excellent piece of research writing. It is based on the final report of a research project sponsored by the University Grants Commission, New Delhi. While introducing the research problems Pandey has presented an exhaustive and critical review of research articles published by numerous social scientists. This review in itself is a valuable picture of research scenario of child abuse. She has selected a fairly large number of problems spelled out as her research objectives while selecting relevant variables from antecedent conditions as well as from the consequent conditions of child abuse, she has made an exhaustive list of them. As a matter of fact she has included in her list all the important variables investigated by early researchers. She had a few more important as well as critical consequences of child abuse for the psychological development of abused children. It is self-evident that she has successfully completed a herculean task under strict temporal and financial constraints on the required extensive fieldwork.

In order to measure the diverse facets of antecedent conditions and consequent psychological consequences a number of tools developed and standardized by some eminent psychologists, along with some developed by the author herself were employed. These tools generated enormous mass of data. Notwithstanding the fact that analysis of results contain some infirmities, but no analysis of results are perfect. Whatever analyses have been made has yielded interesting facts and they lead to fresh insight. The reported results are enough to meet the requirement of the research objectives. It is also correct to asseverate that more sophisticated and intensive analyses would have yielded more fruitful insights about the dynamics of child abuse. However, the mass of data generated are virtually a gold mine for further analysis and generalization.

In the discussion part of the monograph the author has thoroughly discussed the results and brought their implication with adequate sagacity. The most interesting and fresh path paving observations have emerged. In writing about various shortcomings and limitations of her methodology and data analyses along with theorization reveals sincerity and integrity of the researcher. These portions pave the way for developing new strategies for further investigation of the dynamics of child abuse. I am quite confident that this monograph will prove to be a landmark in the corpus of child abuse studies and prove useful for designing research projects on other social problems in India.

L.B.Tripathi
Former Head
Department of Psychology
D.D.U. Gorakhpur University
Gorakhpur-273009

HIG-32, Pritam Nagar,
Dhumanganj,
Allahabad-211011

Preface

Over the last few decades, the incidences of child abuse and neglect have increased dramatically at an epidemic rate and have challenged the existence of child population. Primarily, child abuse emerged as a social indicator in developed countries (Kempe *et al.*, 1962) and later on, it was realized that it exists in same form or other throughout the world and has reached to an alarming volume. Although preventive services were proliferated in developed countries but no significant attempt has been made to understand the detrimental causes and consequences, both social and psychological, of child abuse and neglect in developing countries like India.

The problem struck to my mind a few years ago, when I was engaged with fieldwork, for my earlier research project on "Psycho-Social Correlates of Population Control and Reproductive Health". I observed that in some communities, parents preferred to have larger number of children for economic support. Despite admitting their children to school, parents used to send them for earning in the field, tea stalls, industries, or household work. Preferably, girls were found more abused and neglected than boys within and outside the family.

Such events were eye-opener and oriented me to plan for this study to examine, "social dynamics of child abuse and its psychological consequences in eastern districts of Uttar Pradesh". The data were collected from rural-urban settings of Gorakhpur, Deoria and Maharajganj districts, under Gorakhpur division. These districts are densely populated and underdeveloped and various forms of abused cases are reported frequently. Surprisingly, this problem is not limited to poor/low status families but is also found in same form or other in other strata of society.

The study investigated the influence of socio-demographic and personal factors on child abuse and neglect. Further, present researcher attempted to explore the consequences of child abuse on cognitive, motivational, and behavioural processes as well as personality development of children. Findings evinced numerous surprising facts.

This monograph entitled, *Psychological Consequences of Child Abuse*, consists of four major chapters. In the first chapter, the incidence and prevalence rates of child abuse in India and abroad, causal factors and consequences of abuse on development of children have been systematically reported and the objective and hypothesis have been derived on the basis of pertinent literature mentioned in this section. Chapter 2 focuses on plans, strategy and procedure of the study. The statistical treatment of data, and obtained results have been recorded and also presented graphically and in tabular forms in Chapter 3. In the last chapter, findings of present study have been interpreted and discussed with the help of numerous empirical and theoretical evidences of child abuse and its outcomes.

Further, a theoretical model has been developed on the basis of present findings. Although, the author is aware of the limitations of this piece of research and does not claim that this phenomenon has been thoroughly explored, yet, undoubtedly, she can assert that this research is a pioneering effort to investigate the role of socio-psychological processes in the origin, prevalence and outcomes of child abuse and neglect. Results of the study suggest the need for preventive interventions to both, abusive parents as well as children, who are victims of various forms of maltreatments. The researcher seriously feels the need for developing innovative preventive programmes to educate parents/adults about the adverse consequences of abusive acts on the development of their children. Thus, the present piece of research has raised many more issues, which are still unresolved. However, this study will provide base to investigate these problems in further researches.

Dr. (Mrs.) Sushma Pandey

ACKNOWLEDGEMENTS

It is with a great sense of gratitude that I acknowledge the guidance and support, which I received from inception to completion of this research monograph. Since this investigation was carried out with the financial assistance provided by the University Grants Commission, New Delhi, thus, I express my sincere gratitude to the organization for timely support to complete the project without difficulty.

During the conduction of this study, I consulted and received guidance, support and encouragements from a number of renowned senior psychologists. First of all, I express my sincere gratitude to Prof. L. B. Tripathi, Former Head, deptt. of psychology, D.S.A, D.D.U. Gorakhpur University for his persistent support, invaluable suggestions and guidance throughout the period.

I also express my sincere gratitude to Prof. Arun Kumar, Vice Chanceller, D.D.U. Gorakhpur University, and (Mrs.) Meena Kumar for their invaluable suggestions, support and encouragements.

I am also grateful to Prof. (Mrs.) A. Agarwal, Head, deptt. of psychology, D.S.A., D.D.U. Gorakhpur University, for her valuable suggestions and encouragements.

I also consulted and received suggestions and support from many academicians of other universities amongst whom the names of Prof. R.C. Tripathi, Director, G.B. Pant Social Science Institute, Allahabad, Prof. K. Mahadevan, Former Head and Dean, S.V. University, Tirupati, Prof. R.C. Mishra, of B.H.U., Prof. G. Misra, and Prof. Anand Prakash, of Delhi University, Prof. S.N. Rai, of Meerut University, Prof. P. C. Misra, of Lucknow University, Dr. K. N. Tripathi of Bhopal University and Prof. Herumb Chaturvedi of Allahabad University , need special gratitudes.

I extend my heartfelt thanks to my husband Dr. R. N. Pandey, Reader in Economics, and to my parents, brothers and children (Avanish & Rishabh). They have been constant sources of motivation and pillars of support throughout the period.

Last but not the least; I must be thankful to all the parents and children who participated in this investigation as participants. I extend thanks to my students who worked silently with great devotion, throughout the period. My thanks are also due to Dr. Saurabh Kr. Tiwari and Mr. Pragyesh Kr. Mishra for computing, typing the report and transforming it into the presentable form.

Dr. (Mrs.) Sushma Pandey

Contents

List of Tables

List of Figures

1
INTRODUCTION

Child abuse is a social psychological phenomenon, which is present at epidemic rate in same or other forms throughout the world. The problems emancipating from child abuse and neglect are indeed complex and multifaceted. Primarily, child abuse emerged as a social issue in developed countries. Later on, it was realized that it is present all-over the world and has achieved alarming volume. In 1974 the first National Incidence Survey (NIS) was conducted in United States of America and 60,000 cases of child abuse were recognized. Again second NIS survey conducted in 1980, reported 1.1 million cases of abuse whereas the 1989 survey (NIS III) presented 2.4 million figure of abuse and recently, the 1997 NIS survey identified the statistics of 9,69,000 figures of child maltreatment. Over the last few decades, as the magnitude and severity of the problem have become apparent, the preventive intervention programmes have proliferated in developed countries. Recently, the international organizations have started launching programmes to generate international awareness about the root causes and consequences of child abuse on overall development of children (Taylor, 1991; Finkelhor & Korbin, 1988; Korbin, 1987; Rohner & Rohner, 1980).

Child abuse is neither a recent phenomenon, nor it is emerged with any historical period or cultural practice. It is as old as the history of mankind. However, this problem gathered the attention of scientists and practitioners, following a seminal paper on the battered child syndrome in early 1960s, presented by Henry Kempe *et al.*(1962). Later on, U.S. government called it "a National emergency". As a result, the legislation, clearly defining and mandating the reporting of child abusc and neglect, was at once introduced in the form of Child Abuse Prevention and Treatment Act, i.e. P.L. 93-247, passed in 1974. This Act defined child abuse as, "the physical or mental injury, sexual abuse, negligence or maltreatment of the child under the age of 18 by a person who is responsible for child's welfare under the circumstances which indicate that the child's health and welfare is threatened thereby" (Child Prevention and Treatment Act; P.L. 93-247, 1974). Though, this law was vague in practical purposes thus, further amendments were made in 1988, known as "Child Abuse Prevention, Adoption and Family Services Act (P.L. 100-294, 1988)". Besides all the efforts, the statistics of abuse and neglect are increasing at alarming rate. The National Incidence Survey (NIS) reports of 1980, 1989, 1993 and 1997 indicate the increasing proportion and severity of the problem during the past few decades. Apart from numerous efforts made at national and international levels, this problem is still remaining a serious psycho-social issue, existing at alarming volume throughout the world in general and in developing countries in particular. In Indian society also the child abuse and neglect are present at epidemic threshold, but the exact number is almost unknown due to the lack of reporting statistics and biases.

SOCIAL DYNAMICS OF CHILD ABUSE AND NEGLECT

The social dynamics of child abuse and neglect include various contextual, socio-cultural and familial factors closest to the child's social world. Researchers (Cicchetti & Olsen, 1990) pointed out that in order to understand the developmental significance of child maltreatment, one has to recognize how the parent's failure to provide nurturing, sensitive and supportive care giving, makes any form of maltreatment, particularly harmful to child development.

Since children are the most vulnerable segment of human population, who are primarily dependent on adults (parents/elderly caretakers). They represent one fourth (1/4) of the world population. But they are not equipped to protect themselves and bound to confront with several problems generated in the family and society. Historically, corporal punishment was the well-accepted rule of disciplining children till 20th century. However, attitude towards children and method of disciplining them varied with the passage of time. Present century brought not only changes in child rearing practices and treatment with legal protection for the child's right but also raised quest for all preconceptions about children and their development. Despite numerous efforts made by academicians, professionals, government/legislation as well as non-government organizations, adults' treatment with children are not yet satisfactory and often they (children) become victims of abuse and exploitation. Unfortunately, millions of children in India are living in a pathetic condition and bound to survive with disgraceful and dehumanized lifestyle (Baruah, 2003).

The status of children in India is no way better than worldwide position. The harsh reality is that a large segment of child population is prey to various forms of child abuse and neglect. In India, the incidences of child abuse and neglect are growing at epidemic rate and such horrifying realities of child abuse, challenges the defining sentence in the National Policy for Children; "Children are supremely important assets and their nurture and solitude are the nation's primary responsibility" (National Policy for Children, 1974).

In fact, most of the societies including Indian society have long ignored and treated child abuse as part of discipline. Hence, domestic violence has not been recognized as a serious social problem. It is perceived as a private family matter and not opens to discussion in front of outsider. From socio-psychological perspectives, questions raised are : whether domestic violence is a private issue or a matter of public interest? whether parental violence against their children should be regarded as disciplining or would be considered as a form of child (physical) abuse? Under what conditions, parental violence is viewed beneficial or become abuse and neglect? Such issues in Indian context should be discussed in relation to multiple causative factors. Although an overwhelming majority of such cases are unreported, thus, actual statistics of this problem is unknown. Cases that come to the knowledge of official agencies—hospitals, schools, nursing homes, police and court records etc. are very few and appear as tip of the iceberg. Yet researchers should identify the cases of abuse and neglect within the family on the basis of authentic measures.

Conceptualization of Child Abuse and Neglect (Maltreatment)

Child abuse and neglect are understood as complex social psychological phenomena. However, controversy exists in defining child abuse and neglect (maltreatment) because of multiple reasons. Since the nature and forms of child maltreatment vary from culture-to-culture and nation-to-nation. Thus, defining child maltreatment is obviously problematic because these definitions are subjective and constantly changing. Despite several disagreements, definitions are the essential part of any research. Apart from this, child protection programmes or schemes are largely depends on understanding of definitions. Yet, attempts to define and clarify the concept of child abuse runs into several difficulties. Firstly, the concept has largely been clarified in western context where it has been extensively studied. Secondly, there is lack of clarity and specificity in use of the term. To some extent, this issue is multidisciplinary; consequently medical researchers refer to "child battering" who focuses on the need for intervention whereas; psychiatrists/psychologists to "child abuse" which emphasizes emotional aspect of the situation. Quite often, the terms "child battering", "child neglect", "child abuse", "child maltreatment" and "physical abuse", are used by different authors to refer to the same situation. "Child maltreatment is a generic phrase that covers a wide range of behaviour. It covers all behaviour patterns, whether or not sanctioned by law and custom, which are in some way injurious to the Child's health or social, economic, emotional or moral wellbeing." It comprises of a wide range of behaviour distinguished conceptually as well as operationally. For analytical purpose, such behaviours have been grouped under broad categories. It is to be noted that these are not mutually exclusive categories. Social scientists, psychologists, child advocates and professionals, sincerely working in the areas of child maltreatment have defined this issue in different manner.

Legally based definition heavily focuses on evidentiary criteria in an effort to prove or disprove an act of abuse. Therefore, The Child Abuse Prevention and Treatment Act, i.e. P.L. 93-247, passed in 1974, defined child abuse as, "the physical or mental injury, sexual abuse, negligence or maltreatment of the child under the age of 18, by a person who is responsible for child's welfare under the circumstances which indicate that the child's health and welfare is threatened thereby." (Child Prevention and Treatment Act (P.L. 93-247, 1974). However, to identify the nature of child abuse a widely accepted definition of child abuse and neglect is barely essential, involving two main objectives : Firstly, it should distinguish child abuse from other social, economic and health problems of international concerns and secondly, it should be sufficiently flexible to apply to widerange of child abuse situations in a variety of social and cultural contexts. Taking these perspectives into consideration, Finkelhor and Korbin (1988) have defined child abuse, "as the portion of harm to children that results from human action which is proscribed, proximate and preventable". The term 'proscribed' refers to any human action concerning harm, if it is evaluated negatively with its deviance, harmful intent, violation of law and social expectation. Such action has to be 'proximate' which refers to the action of immediate caretakers closer to child's immediate environment and the term 'preventable' denotes that if some alternative course of action is potentially available in the setting but the caretakers to check harm do not exercise it, such actions are also considered abuse.

Thus, lack of consensus in the definitions of child abuse is witnessed. Legal professionals focus on documentation of abuse and proof of the occurrence of abusive acts by perpetrators. Contrary to this, researchers imply variety of definitions depending on their theoretical perspectives of research requirements. However, professionals have given relatively broader definition and focus on the injuries and sufferings that result from child maltreatment.

Forms and Patterns of Abuse

A sizeable number of researchers and professionals working sincerely in the field of child maltreatment have reported that the forms of abuse and neglect differ nation to nation yet there is consensus on some forms or patterns of abuse, which are found throughout the world wolfe, 1999; (Finkelhor, 1996; Korbin, 1987; Straus & Steinmetz, 1980). It is classified into four major types, viz. Physical abuse, Neglect, Sexual Abuse and Psychological or Emotional Abuse (Wolfe, 1999; Straus, 1980). On the basis of three National Incidences Studies (NIS) conducted by the U.S. Department of Health and Human Services in 1981; 1988 and 1993, the definitions are framed following two different sets of definitional standards: (i) the harm standard, whereby child has suffered from apparent harm as a result of maltreatment as sign of cuts, belts or broken bones, and (ii) the endangerment standard, which includes entire forms of harm, adding those maltreatment experiences, put them in danger of being harmed, such as witnessing violence in the family (Sedlak & Broadhurst, 1996).

1. *Child Physical Abuse (CPA)*

Child physical abuse refers to the inflection or endangerment of physical injury due to beating, kicking, hitting, burning or otherwise harming a child. Sometimes injuries are intentional (parents' intention to control the child), but nonetheless corporal punishment and harsh disciplinary tactics can result in child injuries. The nature of injuries, resulting from physical abuse may range from minor (bruises) to moderate (scars) to severe (burns, sprains or broken bones etc.). However, physical injury denotes only the visible scars of abuse. Due to harsh and insensitive treatment, physically abused children are often found destructive and aggressive than normal children and many show major emotional and cognitive problems (Aber & Cicchetti, 1984).

2. *Child Neglect (CN)*

Physical and emotional neglects are characterized by inability of caregivers to provide children's basic physical, educational or emotional needs :

(i) *Child Physical Neglect (CPN):* Physical neglect covers refusal of or delay in seeking health care, abandonment, expulsion from the home or refusal to allow a runaway to return home and inadequate supervison.

(ii) *Child Educational Neglect (C Edu. N):* Practices such as allowing chronic truancy, failure to admit a child at appropriate school age in the school and failure to provide special educational requirements or sending child for wages instead of attending the school etc. are considered educational neglect.

(iii) *Child Emotional Neglect (CEN):* It includes actions such as marked inattention to the child's needs for attachment, or affection, failure or refusal to give psychological care, spousal abuse in front of the child and permission of drug or alcohol use by the child. However, the determination of neglect requires the consideration of cultural values and contextual factors embedded in poverty and deprivations.

3. *Child Sexual Abuse (CSA)*

This form of child abuse involves fondling a child's genitals, intercourse, incest, rape, sodomy, exhibition and commercial exploitation through prostitution or the production of pornographic materials etc. Sexual abuse may be underreported because of the secrecy or conspiracy of silence that so often characterizes these cases. The behaviour and development of sexually abused children may be affected significantly, especially in relation to longer duration/greater frequency of abuse, the use of force, penetration and a closer relationship to the perpetrator (Kendall-Tackett, Williams & Finkelhor, 1993; Finkelhor & Browne, 1988).

Unlike physical abuse and neglect, sexual abuse has no connection to child rearing, discipline, or inattention to developmental needs. Rather, it constitutes a breach of trust, deception, intrusion and exploitation of a child's innocence and status (Finkelhor, 1995). Whereas all types of maltreatment share a common ground in relation to the abuse of power by an adult over a child. Sexual abuse stands out from physical abuse and neglect in term of these specific dynamics.

4. *Child Psychological Abuse (C Psy A)*

Psychological maltreatment of children consists of acts of omission and commission, which are judged on the basis of combination of community standards and professional expertise to be psychologically damaging. Individuals commit such acts singly or collectively, who are in a position of differential power (e.g., age, status, knowledge and organizational form) that render a child vulnerable to abusive practices. Such acts damage immediately or ultimately the behavioural, cognitive, affective or physical functioning of the child. Examples of "psychological maltreatment includes acts of rejecting, terrorizing, isolating, exploiting and miss socializing."

O'Hagan (1995) has argued that psychological abuse and emotional abuse are not the same, although they are not entirely separate experiences and has made the distinction that emotional abuse affects emotional life and development, whereas psychological abuse affects mental (i.e. cognitive) life and development. The term of psychological maltreatment, however, has widespread support among theorists and researchers for the advantages of interventions and researches (APSAC, 1995).

The APSAC Guidelines (1995) include the following widely accepted broad conceptual statement and psychological maltreatment categories. "Psychological maltreatment" means "a repeated pattern of caregiver behaviour or extreme incident (S) that conveys to children that they are worthless, flawed, unwanted, endangered, or only of value in meeting another's needs". Psychological maltreatment includes (a) spurning, (b) terrorizing, (c) isolating, (d) exploiting/ corrupting, (e) denying emotional responsiveness, and (f) mental health, medical, and educational neglect (APSAC, 1995).

Epidemiology of Child Abuse and Neglect

Since child maltreatment is widely studied problem in developed countries, thus accurate figure is nearly known. National Incidence Study (NIS) reports that almost 3 million children are harmed or endangered by abuse or neglect each year in the United States amounting to a total yearly incidence rate 42 of every 1,000 children (Sedlak & Broadhurst, 1996). This rate is derived from estimates of National Incidence Study (I, II, III, & IV) based on official reports of all four types of maltreatment as well as from a nationally representative sample of professionals who came in contact with maltreated children in a variety of settings (Figure 1.1).

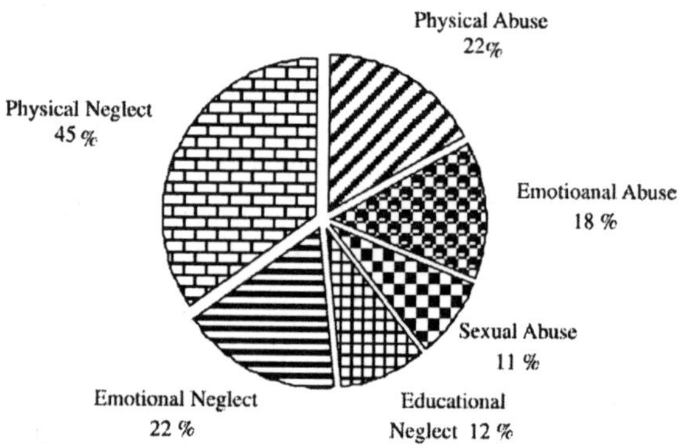

Fig. 1.1: Displays the percentage of each sub-types of child maltreatment (based on NIS-III report).

It is clear from Figure 1.1 that the percentage of reported cases of maltreatment broken down according to each sub-types, proportions that has remained fairly constant over the past decade. Child neglect continues to hold the title as the most commonly reported from of maltreatment, affecting almost 30 children of every 1,000 in the United States and accounting for 70 per cent of all reported incidents. Physical, Sexual and Emotional abuses affect another 18 children per 1000, or 43 per cent of the total (these percentages exceed 100 per cent due to overlap). Specifically, physical abuse accounts for 22 per cent, sexual abuse for 11 per cent and emotional abuse for 18 per cent of maltreatment incidents (again, with overlap in reports and substantiated case rose dramatically in the 1980's but have since levelled off) (Sedlak & Broadhurst, 1996).

Further, official incidence rates provide a useful comparison of reported rates of child maltreatment over time, but they likely underestimate the actual prevalence of abuse by parents throughout the country (Straus & Kantor, 1994). Apart from this population-based estimate, which is also subject to reporting bias presents the incidence of physical abuse 5 times higher than the NIS-based rate of physical, sexual and emotional abuses combined. A third approach to estimating the extent of the problem uses lifetime prevalence estimates of maltreatment as a child. Males (32.2 %) reported more child physical abuse than females (21.1 %), while sexual abuse during childhood was more commonly reported by females (12.8 %) than males (4.3 %). Retrospective reports of childhood maltreatment are also imperfect but current evidence suggests valid and reliable estimates of significant past events (Bernstein *et al.*, 1994).

The Increasing Rate of Child Maltreatment

It is difficult to ignore the finding that the total number of abused and neglected children reported to authorities has nearly doubled since 1986. Physical abuse nearly doubled, sexual abuse more than doubled, emotional abuse, physical neglect and emotional neglect reports are almost more than 2.5 times greater. Moreover, the total number of children seriously injured and the total of those endangered by maltreatment have quadrupled since 1986.

These dramatic increases in estimated number of maltreated children prompt the question as to whether the estimates detect an actual increase in maltreatment or merely reflect a greater awareness and reporting than previous generations. NIS investigators believe that both awareness and actual prevalence have increased.

Child Abuse and Neglect in India

The incidence of child abuse is increasing day by day and has achieved an alarming threshold in India. However, exact figure is unknown due to reporting biases. According to an estimate of the Juvenile court (1980-81), the largest number of maltreated children were reported from Maharashtra (8016), Karnataka (1572) and Delhi (1246). Mehta (1979) reported that in Bombay, among the 150 cases of rape victims studied, the youngest was a six month old child. Although to check the epidemic proportion of child abuse, legal efforts are made at national and international levels, yet the rates of abusers and abused are increasing. It is highly prevalent in poor community but is also present in some forms or other in different strata of society.

Despite this, the available official statistics on maltreatment against children (as annually reported cases of child maltreatment in India) do not indicate the actual extent of the criminal victimization of children owing to large number of cases remaining unreported or unregistered. These statistics, however, provide us with a peep into nature of crimes, which come to the knowledge of the police. A report (1997) of crime against children in India tells us that amongst the 10 types of crime against children, the child sexual maltreatment tops the list followed in descending order by kidnapping and abduction, procreations of minor girls, selling of girls for prostitution, buying of girls for prostitution; abetment of suicide, exposure and abandonment, infanticide, feticide and marriage of minor children.

Child rape (up to 16 years) has shown noticeable figure in its reported incidence from 1993 to 1997 and percentage variation has been 8.1 per cent over 1993 figure. Almost similar increase was noticed in other crime heads. A further probe into the rising incidence of child rape revealed that (i) of all the victims in rape cases during 1997, children alone accounted for 28.8 per cent. (ii) Since 1993, the cases of rape of children below the age 10 years were on a rise every year in the country except in 1996. (iii) For the first time in the last 5 years, the year 1997 witnessed a sharp rise of 26.6 percent, a rise of 4 per cent in victims in the age group of 10-16 years; and (iv) the child rape victims below 10 years reported still increase in 1997 over 1996. Since in Indian studies most of the reported cases of child abuse are identified from organized sectors in the form of child labour, beggars or street children, however, a sizeable number of such cases from unorganized sectors in the form of domestic servants, victims in orphan houses or residing at railway platforms and victims in the families are seen in plenty. Unfortunately, such cases are unreported, and unnoticed, though, ruining the most vulnerable segment of population. Since child abuse is a serious socio-psychological problem, therefore, intensive researches and preventive efforts should be made at multiple levels. Unfortunately very little efforts have been made in this area.

RESEARCH WORK DONE ON CHILD ABUSE IN INDIA

Since dynamics of child abuse are multifaceted, thus, multidisciplinary research efforts are needed to understand the root causes, consequences and strategies for prevention of this evil. Researchers reported that child abuse occurs more frequently during rapid socio-economic changes, urban migration and family disorganization (Chattoraj & Saxena,1990; Korbin, 1987; Mehta, 1979). Some of the deep-rooted reasons of child abuse and neglect identified in India are: population density, poverty, unemployment, lack of education, social isolation, high mobility, distorted family relationship, poor bond between parents and children, negative parenting style, high anxiety and stress, poor coping ability, traumatic disastrous experiences, childhood abusive experiences and so on (Pandey, 2005; 2004; Baruah, 2003). Mehta (1979) identified that motive behind kidnapping of children was mainly prostitution, selling, sexual gratification, unemployment, extreme poverty, broken homes, anti-social surroundings, ignorance and illiteracy. In another study, Bhattacharya (1979) showed that children are generally the victims of prolonged and severe malnutrition, under nutrition resulting impairment and underdevelopment or growth. Whereas, Faller (1981) indicated that the environmental, familial and individual parental factors and their interplays are often the causes of abuse and neglect of children.

On the basis of an extensive study, Gil (1971) reported that stress and frustration are the major causal factors of child abuse. Stress caused by deprivation or poverty, high density in over crowded housing, large number of children, single parent families, the absence of child caregivers and poor education were found to be associated with abuse and neglect. Pandey (2004; 2002) identified that high anxiety, stress, poor family environment and negative parenting styles were found positively associated with child abuse and neglect.

Legal Efforts

Although, recognizing the vulnerability of child and their significance as future citizens, the Constitution of India gives a special place to protection of children from all forms of exploitation and abuse. Whereas, Article 39 provides that children at a minor age are not to be abused but protected against exploitation and against moral and material abandonment. Article 25 prevents the employment of children in hazardous occupations. The National Policy for Children (1974) made efforts to provide protection and care for them. Despite these legal efforts, no single legislation deals exclusively with the problem of child abuse and neglect in India, however, there are numerous central and state laws, which have been enacted with a consideration on child exploitation, cruelty and neglect. Similarly, the Indian Penal Code (Sec. 363A) deplores acts like kidnapping or maiming the child for the sake of begging, selling or for the purpose of prostitution (Sec. 372-373), exposure and abandonment of the child under 12 years by parents (Sec 317), is viewed as serious offences. The Juvenile Justice Act (1986) provides for the "care, protection, treatment, development and rehabilitation of neglected or delinquent Juveniles (Juvenile justice Act, 1986).

Cruelty and wilful neglect of children by their parents or caretakers in any form is punishable, under sub-section (1) of Section 41, imprisonment up to six months or with fine or both. In Section 42, employment of juveniles for the purpose of begging is also a punishable offence. Under Section 43 of the Juvenile Justice Act, giving intoxicating of juvenile employees, Section 50 provides authority to a police officer to take charge without warrant of a juvenile who has escaped from a special home to send him back there. Despite constitutional guarantee and legislative measures, it is found that a large number of children are still the victims of various forms of maltreatments.

A cursory glance at the welfare statistics on children, however, denotes only a part of the reality, which is quite different. Socio-psychological researches show that the children have been admitted to the institution due to abandonment, parental neglect, broken homes, undesirable home environment, and delinquency, running away from home or due to their being lost.

The cases in which children are abused and neglected are grouped under two broad sections: (i) Crimes committed against children who are punishable under Indian Penal Code (IPC) and (ii) Crimes committed against children who are punishable under special and local laws :

(i) The crimes which are punishable under the Indian Penal Code include: (i) foeticide (crime against unborn child), Sections 315 and 316 IPC; (ii) infanticide (crime against newly born child), Section 315 IPC; (iii) exposure and abandonment (crime against children by parents or other to expose or leave them with the intention of abandonment), Section 317 IPC; (iv) kidnapping or abduction for exporting, Section 360 IPC;

kidnapping for ransom, Section 363 read with Section 384 IPC; for camel racing, Section 363 IPC; for begging, section 363-A IPC; to compel for marriage, Section 366 IPC and for slavery, Section 367 IPC; (v) procurement of minor girls (for inducement to force or seduce to elicit intercourse), Section 366-A IPC; (vi) selling of girls for prostitution, Section 372 IPC; (vii) buying of girls for prostitution, Section 373, IPC and so on.

(ii) Crimes against children, which are punishable under special and local laws, fall under the ambit of: (i) Immoral Trafficking Prevention Act, 1956; (ii) Child Marriage Restraint Act, 1979; (iii) Child Labor Prohibition and Regulation Act, 1986; and (iv) Juvenile Justice Act, 1986.

Contributing Factors of Child Maltreatment

Since child maltreatment is a psycho-social problem and its causal roots are multifaceted. Although earlier 1960s studies indicated that this problem is rooted in adult's psychological disturbances. Abusive adults were found as abused in their own childhood, having unrealistic expectation that children would satisfy their own unmet emotional needs, and poor aggression control (Spinneta & Rigler, 1972; Kempe, *et al.* 1962), yet recent studies did not confirm "abusive personality type" variable. Certain other parental and child characteristics of parents include: disturbances, substance abuse, history of abuse as a child, belief in harsh physical discipline, desire to satisfy unmet emotional needs through the child, unreasonable child behaviour, young age (below 30 yrs) and low educational level. A close association of parental factors, i.e., poor parenting skills, mentally disturbed parents, parental stress, drug abuse with child maltreatment has been identified. Certain child characteristics prone to being target of abuse are: premature or very sick baby; children who are inattentive and overactive and having difficult temperament and other developmental problems. Current approaches to understanding child abuse and neglect emphasize an interlinked cluster of individual, family, environmental and socio-cultural risk factors that can result in increased risk of maltreatment, as well as the possible protective mechanisms that disrupt this etiological process (National Research Council, 1993). Societal and cultural factors that influence the formation of a healthy parent-child relationship play prominent roles in child abuse, in addition to specific psychological processes affecting the dynamics of the relationship.

Belsky (1980) presented a conceptual framework for integrating diverse theories on etiology of child maltreatment (Table 1.1).

Table 1.1: The causes of child maltreatment

Ontogenetic Factors	*Micro System Factors*	*Exo-system Factors*	*Macro System Factors*
History of abuse	Marital discord	Inadequate health care facilities	Economic recession
Alcohol abuse	Single parenthood	Social isolation	Cultural acceptance of corporal punishment
Stressful experiences	Premature or unhealthy child	Unsafe neighbourhood	View of children as possessions
Low I.Q.			
Psychiatric and physical illnesses			

This framework includes psychological disturbances in parents, abuse eliciting characteristics of children, dysfunctional patterns of family interaction, stress-inducing social forces, and abuse promoting cultural values. This model is mainly based on Bronferbrenner's (1977, 1979) work related to the ecology of human development which conceptualized child maltreatment as a "Social Psychological Phenomenon" and multiple forces determine it across four different levels: in the individual (Ontogenetic development), the family (the micro-system), the community (the exo-system), and the culture (the macro-system), in which both the individual and family are embedded. Belsky combined Tinbergen's concern for ontogenetic development with Bronferbrenner's concern for the ecology (or the context) for developing his conceptual model of child maltreatment. He assumed that forces within the individual determine child maltreatment; the family, the culture and these determinants are nested ecologically with one another. Thus, causes of child maltreatment cannot be explained on the basis of single factor. It is the product of multiple factors interlocked with each other. Despite the parent characteristics, several interaction variables at the family, community, society and culture play prominent roles.

Table 1.2: Displays the predictors of child abuse and neglect

Factor	*Description*
(i) Parent / Adult Characteristics	Psychological disturbance; substance abuse; history of abuse as a child; belief in harsh physical discipline; desire to satisfy unmet emotional needs through the child; unreasonable expectations for child behaviour; young age (most under 30); low educational level; cognitive deficits in abuser, negative affect etc.
(ii) Child Characteristics	Premature or very sick baby; difficult temperament; inattentiveness and over activity; other developmental problems, child behaviour etc.
(iii) Family Characteristics	Low income; poverty; homelessness; marital instability; social isolation; physical abuse of mother by husband, frequent moves; large closely spaced families; overcrowded living condition; disorganized household, lack of steady employment; other signs of high life stress etc.
(iv) Community	Characterized by social isolation, few parks, day-care centres, preschool programmes, recreation centres and churches to serve as family supports etc.
(v) Culture	Approval of physical force and violence as ways to solve problems, socio-cultural influences, poverty and social disadvantage and corporal punishment as disciplinary practices etc.

Parent/Adult Characteristics

Major parent/adult characteristics associated with child abuse and neglect are: role of parenting, childhood experiences, stress and coping abilities, distorted beliefs, child rearing abilities and resources, and emotional reactivity. Belsky (1984) suggested that the quality of parenting is multiply determined by characteristics of parent, the child and social context.

1. *Early Maltreatment Experiences as a Child*

Researchers endeavored to identify relationship between childhood experiences of abuse and abusive treatment (Kempe & Helfer, 1972). Since prior abuse and family violence in childhood may perpetuate a cycle of violence across generation, this issue remains an important, yet not fully resolved, concern. Abusive background significantly increase the likelihood that an individual will become subsequently involved in coercive relationship with peers (Dodge, Bates & Pettit, 1990) dating partners (Wolfe *et al.*, 1998; Malamulh *et al.*, 1991), and their own children (Straus, Gelles & Steinmetz, 1980; Gaines *et al.*, 1978).

Maltreating parents usually have experienced psychological deprivation or maltreatment as children and suffer persisting conflicts about care and control. In adult life, they form relationship in which these conflicts are reenacted through excessive, but often ambivalent, dependency. They often look to their children for emotional comfort giving priority to their own need instead of their children. Demands for care taking from adults or infants readily provoke the parent's anxiety and knowledge.

2. *Cognitive Deficits*

Child abuse seldom involves isolated or premeditated events but rather occurs in the context of child rearing. Studies of the everyday interactions between abusive parents and their children, accompanied by knowledge of what abusive parents are thinking and feeling during these interactions have afforded a useful understanding of the nature of these overt and covert processes. Numerous studies identified a strong association between knowledge of child development and potential for abuse (Larrance & Twentyman, 1983). Milner and Wimberly (1980) underscored the excessive rigidity in expectations of parents who are high in child abuse potential. Other studies (Newberger & Cook, 1983) argue that abusive parents were less able to assess the developmental needs of their children. Furthermore, researchers (Larrance & Twentyman, 1983) reported that abusive mothers were less knowledgeable about the capabilities of their children. Whereas, other studies suggest that abusive parents tend to have a limited repertoire of parenting skills as well as inappropriate expectation for child behaviour. Trickett and Susman (1988) reported that abusive parents were found less satisfied with their children and perceived child rearing to be more difficult than the control parents. Researches thus indicate a strong link between parenting skill, knowledge of child development and potential for maltreatment. Wolfe (1985) concluded that abusive parents express significantly less positive behaviour during interactions with their children and other family members. Relative to other families, abusive family members interact less often with each other. When they do interact, the interaction tends to be more negative and involve excessive forms of verbal and physical control that exceed demands of the situation. Earlier researches (Hurlock, 1984, Hollowtiz & Stulberg, 1959) have also denoted a vicious cycle of bad-parent child relationship that becomes increasingly worse as time passes. The deterioration in parent-child relationship continues as it weakens emotional bond between family members.

Cognitive deficits refer to an absence of thinking whether it would be beneficial, whereas cognitive distortion refers to persons who typically do not lack the ability to organize or process information, but their thinking is described as biased, dysfunctional, or misguided (Kendall, 1993). Distorted or negative perceptions that relate specifically to the parent's own experiences and/or perceptions of the child-rearing role may reflect the developmental immaturity, first noticed among sample of abusive parents (Kempe & Helfer, 1972), which reflects the extent to which they accurately recognize the demands and responsibilities that accompany child rearing. Deficits and distortions both may play role in the process of perceptions and judgment of their child's behaviour. Abusive parents may misperceive child behaviour in way that lead to inappropriate responses and increased aggression (Azar, 1997; Milner, 1995; Crittenden, 1993; Larrance & Twentyman, 1983).

3. *Negative Affects (Anger and Arousal)*

Researchers have consistently described abusive parents as impulsive and exhibiting low frustration tolerance. The act of interpersonal aggression is highly attributable to situational cues and characteristics of the individual (Averill, 1983; Berkowitz, 1983; Zillman, 1979). Certain situational cues often involve aversive behaviour or child features and individual characteristics include factors such as over sensitivity (Knutson, 1978), disinhibition of aggression (Zillman, 1979) and limited interpersonal skill (Burgess, 1985). Physiological arousal (i.e., increased respiration, pulse rate, muscular tension), can lead to aggression; especially if the person labels the source of such arousal as anger provoking (Berkowitz, 1983). Moreover, negative arousal interferes with rational problem solving. This association between arousal and child behaviour may occur gradually during everyday parent-child interaction during highly stressful episodes (Berkowitz, 1983).

Emotional arousal by itself, however, is normal and adaptive reaction to many situations, hence, the development of an abnormal pattern requires an awareness of other adult and situational characteristics that require a multidimensional explanation. Deviant arousal patterns most likely emerge as a function of an individual's predisposition, such as over reactivity and sensitivity to emotional cues and prior experiences (Bower, 1981).

4. *Coping with Stress*

Stress is found in many forms for different individuals and it can be understood as demands that create an acute or chronic imbalance for the individual or the family system. This imbalance, in turn, is typically met by counter pressure from the persons' coping efforts to return himself/herself to a more comfortable state of equilibrium. Since, coping is considered an important resource, which may help, individual to maintain psychosocial adaptation during stressful situation. Researchers have conceptualized coping under three distinct approaches, viz. trait or disposition, contextual approach and transactional approach. In recent researches the last approach is widely exercised with this view, Lazarus and Folkman (1984) define, "coping is an individual's constantly changing cognitive and behavioural efforts to manage specific external and/or internal demands which is appraised as taxing or exceeding the resources of the person", Lazarus (1981) indicated four main functions of coping, viz. (i) to reduce harmful environmental conditions and enhance prospects of recovery, (ii) to tolerate or adjust to negative events and realities, (iii) to maintain a positive self-image, and (iv) to maintain emotional satisfying relationship with others. The moderating role of coping in stress - health relationship, has been thoroughly investigated by a sizeable number of researchers (Pandey & Srivastava, 2004; Misra, 1999; Caplan, 1985).

Child Characteristics

Research issues concerning the child's role in increasing the risk of child abuse have received careful scrutiny because abusive parents often attribute their anger to their child's behaviour (Herrenkohl, Herrenkhol & Egolf, 1983). Children's behaviour, physical feature or developmental limitations may increase the potential for abuse. Child characteristics (age, gender, temperament, low birth weight, hyperactivity, conduct problems, or handicapping condition) have been associated with the risk of maltreatment once environmental and adult factors are controlled (National Research Council, 1993; Whitmore, Kramer & Knutson,1993). Importantly, children are not responsible for their own abuse in any manner; child abuse is an adult action that is never justified. Unintentionally, however, the child's behaviour may still play a role in the continuation or escalation of an abusive or neglectful relationship. Children are emotionally and practically dependent on their parents and therefore, very young children are the most at risk from abuse and neglect. Children born prematurely with physical or intellectual defects, who are poor feeders, suffer illnesses or experience early separation also make extra stressful, emotional demands on their parents.

Demographic Characteristics and Child Maltreatment

Apart from personal characteristics, certain demographic characteristics are also found significantly associated with differential report of child maltreatment (Sedlak & Broadhurst, 1996).

(i) Age Differences

Toddlers, preschoolers and younger adolescents are the most common victims of physical and emotional abuse, which corresponds to the emergence of greater independence and parent-child conflict at both of these developmental periods. Neglect is most often reported when children are quite young. Declining with age, sexual abuse, in contrast, is relatively constant from age 3 onward, which assists to children's vulnerability from early pre-school years throughout childhood (Sedlak & Broadhurst, 1996). Although child physical abuse affects a sizeable proportion of all age groups, the highest rate of physical injury is found among older children (12-17 year of age) probably because of increasing parent-teen conflict.

(ii) Gender Differences

There are few significant gender differences in maltreatment rates with the important exception of sexual abuse. Although girls are

25 per cent more likely to be victims of any form of abuse than boys, this essentially reflects the fact that they account for about 80 per cent of the victims of sexual abuse (Finkelhor, 1993). Moreover the dynamics of sexual abuse differ considerably for boys and girls. Although boys as well as girls are more likely to be abused by someone they know and trust, boys are more likely to be abused by male non-family members whereas girls are more likely to be sexually abused by male family members (Finkelhor, 1995). This finding suggests different patterns of vulnerability for boys and girls, which have implications for safety and prevention.

Family Characteristics

Apart from parental factors, family dynamics are important for explaining the range of psychological sequels of abuse, because it often occurs in the context of multiple functional factors whereas poverty, parental psychopathology, alcoholism and family dysfunctions are prevalent (National Research Council, 1993). Family context is generally described as one of insensitive, marginal parenting and harsh interactions (Friedrich, 1990), even though abusive parents and their children may sometimes behave appropriately under limited or superficial conditions (Crittenden, 1988).

(i) Marital Conflict and Violence

In a study, Straus *et al.*(1980) identified that marital conflict and violence are often associated with child abuse. They estimated that approximately 40 per cent of the families where the adults were violent towards one another; there was also violence towards a child at some point during a 12 month period. Other investigators have reported that in fact marital conflicts and violence most often arise during disagreements over child rearing, discipline and each partner's responsibility in child care (Hilton, 1992; Edleson, Eisikovits, Guttman & Sela-Amit, 1991). Tragically, marital violence and family turmoil not only frighten and disturb children in a direct manner, but the resulting fallout from such events ranging from changes in financial status, loss of family unity and safety, prolongs the stress and harmful impact on children's development (Wolfe *et al.*, 1985).

(ii) Family and Perpetrator

Family demographic characteristics reflect the cultural and social forces that shape child-rearing method at a broad level. Child abuse and neglect appear to be mediated by several major environmental conditions of which social status (SES; family income below the poverty mark, under employment and less education) plays a significant role.

Child maltreatment is found more common among the poor and disadvantaged (Gelles, 1983), especially for child neglect (Drake & Pandey, 1996). It is striking that the incidence rate for all forms of maltreatment rises from 42 per 1,000 (averaged across all SES level) to 99 per 1,000 children among those from the poorest families (earning less than $ 15,000/year). This rate is three times greater than that of children from moderate income families ($ 15,000-$ 30,000/year) and 25 times higher compared with children in families in the highest income bracket (> $ 30,000/year). This finding is not likely due to a reporting bias, because it has remained unchanged for the last 20 years, despite increased awareness and reporting (Pelton, 1994). What does it imply; however, is that the economically based context of maltreatment, viz., restricted child care opportunities, crowded and unsafe housing, and lack of health care and so forth, is powerful contributor to the high incidence rates.

The significance of low SES is felt not only on rates of maltreatment but also growing up with poverty has a substantial effect on the wellbeing of children and adolescents, especially in terms of impairment in learning disability and school achievements (Brooks-Gunn & Duncan, 1997; Schteingart *et al.*, 1995). Moreover, low income is tied to many other forms of disadvantage such as less education, poor paying jobs, inadequate healthcare, single parent status, limited resources, poor nutrition and greater exposure to violence, any of which can impair children's developmental progress significantly (Brooks-Gunn & Duncan, 1997). Poverty has a significant, yet, indirect effect on children's adjustment, most likely because of its association with such negative influences, particularly harassment, inconsistent parenting and elevated exposure to acute and chronic stressor that define the day to day experiences of children in poverty (McLoyd, 1998). Thus, it is widely accepted that lower SES can serve as a significant risk factor to child development and maltreatment, whereas higher SES can serve as an ameliorating protective factor (Masten & Garmezy, 1985).

(iii) Family Structure

Family structure is also connected to the probability of child maltreatment. Children living with a single parent are at significantly greater risk of both physical abuse and neglect most likely because of added stress, fewer resources and opportunities to share child-rearing burdens and lower SES than two parent homes. Children living factors in particular are almost twice as likely to be physically abused as those living with mothers alone (Sedlak & Broadhurst, 1996). Similarly, maltreatment—especially physical and educational neglect— is more common in larger families where additional children in the household mean additional tasks, responsibilities and demands.

Perpetrators of child maltreatment are the child's birth—parents about 80 per cent of the time although there are important exception and key gender differences in relation to different forms of maltreatment. Nearly one half of sexually abused children are abused by persons other than parents or caretakers, compared with only a fraction in other categories. Child neglect is committed predominantly about 90 per cent of the time—by mothers, which fits with the fact that mothers and mother substitutes tend to be primary caretakers and the head of single-parent home. In contrast, sexual abuse is committed more often—about 90 per cent of the time—by males about half of whom are the child's father or father figure. Males are also the offenders in the majority of emotional (63%) and physical abuse (58%) incidents as well (Sedlak & Broadhurst, 1996). Moreover, in 30 per cent to 60 per cent of families where either child maltreatment or woman battering is occurring, one will find that the other forms of violence are also being perpetrated (Edleson, 1999). In their second nationwide survey of domestic violence, Straus and Gelles (1990) found that men who reported beating their wives were also more likely to report abusing their children.

(iv) Social Isolation

Social isolation is one of the commonly noted features of abusive families, coupled with stressful living conditions, such as lack of adequate day care, peer groups or close friends, and adequate housing (Thompson, 1994; Garbarino, 1976). Abusive parents often lack significant social contacts to others in the extended family, the neighbourhood, the community, and to the social agencies most likely to provide needed assistance (Korbin, 1994). These factors play an indirect, yet significant, role in the early formation and healthy establishment of a positive versus abusive parent-child relationship. As a result, abuse and neglect are difficult to detect and community agents who could promote healthy parent-child relationship are less likely to be influential. Apart from this, some cultural factors, such as, child rearing practices, geographical distinctions, and kinship systems might serve as risk-enhancing or risk-reducing factors in abuse and neglect. Social isolation, however, is not a singular factor but rather a set of variables linked to the parent's perception of support and their networks (Coohey, 1996).

(v) Social and Cultural Influences

Many contemporary societies face high rates of child maltreatment in relation to social condition at the community and cultural level (Garbarino, 1997). Researchers have identified a sizeable number of socio-cultural factors, directly associated with child maltreatment.

(a) Poverty and Social Disadvantage

Poverty and social disadvantage are associated with severe restrictions in the child's expectable environment such as lack of adequate day care, safety, and housing, which often impair the development of healthy parent-child relationship collectively, structural determinants of a community's social organization such as economic and family resources, residential instability, household structure, neighbourhood and poverty influence rates of child maltreatment more than any individual or family factor alone (Pelton, 1978; Gil, 1970; Coulton, Korbin, Su & Chow, 1995). In an early study, socio-economic factors accounted for 36 per cent of the variance in rates of child abuse in Pennsylvania (Garbarino, 1976). Unemployment (Light, 1973), restricted educational and occupational opportunities (Gil, 1970), unstable or violent family situations (Straus *et al.*, 1980) and similar disadvantages (e.g., housing, privacy, noise, and pollution levels) often associated with lower social class membership have all emerged as major socio-cultural factors influencing rates of child abuse and other urban problems in North America (Coulton *et al.*, 1995).

Although child maltreatment occurs across all the social strata, there is no doubt that adverse social circumstances, such as poverty, poor housing and ill health, intensify the risk. In addition, maltreating parents are commonly socially isolated, having least contact with their family of origin through geographical mobility, or make little use of social support networks. Despite socioeconomic forces, maltreatment appears to have significant impact on child development above and beyond the influence of stressful socio-economic circumstances alone (Kurtz *et al.*, 1991).

(b) Cross-Cultural Comparisons

Although, it is difficult to draw comparisons between countries because of differences in defining and reporting child maltreatment, what little is known about the prevalence of maltreatment worldwide suggests that unfortunately all forms of child maltreatment are at epidemic proportions in both developed and developing countries. Child abuse is found in all societies and is almost always a highly guarded secret wherever it occurs (World Health Organization, 1997). Studies in which children or young adults were interviewed about their childhood experiences further confirm that prevalence rates of child sexual abuse in other western societies around the world are comparable with North American rates. Clustering around 20 per cent for females and between 3 per cent and 11 per cent for male sexual abuse prior to the age of 18 years (Finkelhor, 1994). (Estimates of child sexual abuse for African Middle Eastern, or for eastern countries are not available). The U.N. also estimates that nearly 2 million children worldwide are being sexually abused, including many who are forced into prostitution and pornography. Similar estimates of physical abuse and neglect unfortunately are not available.

A comparison of incidence rates of child abuse and neglect in the United States, Canada and Australia indicates how the broader social context may influence the incidence and reporting of maltreatment. Canadian estimates extrapolated from the province of Ontario (Trocime, McPhee, Tam & Hay, 1994), suggest that the U.S. has about double than overall rate of maltreatment in Canada (42/1,000 *vs.* 21/1,000). Similar differences are reported for Australia (16/1,000) (Broadbent & Bently, 1997).

(c) Changing Attitude Towards Child Rearing and Corporal Punishment

Child rearing practices have been changing dramatically over the past 50 years. Today's parents are expected to appreciate their child's developmental strengths and limitations and to move away from total reliance on disciplinary control methods towards ones that encourage the child's emerging independence and self-control (Macca by & Martin, 1983). Differentiating child abuse from child discipline remains difficult, because cultural norms in many societies accept corporal punishment as a primary, even necessary component of discipline. Many parents spank their children even though they question its effectiveness (Gallup Organization, 1995). Cultural values, historical precedence and community standards therefore may set the stage for one person's abuse being another person's discipline (Korbin, 1994).

Moderators of Abuse Consequences

Review of studies on child abuse and neglect evinced that the effect of child abuse are moderated by certain factors. However, it does not affect each victim in a consistent manner. Some of the identified moderating factors are discussed below :

1. *Severity and Duration of Abuse*

Severity and duration of the abuse are identified as some of the moderators of abuse. The assumption is that more severe and/or chronic maltreatment will lead to more severe outcomes. Although, empirical data are scarce, some evidence supports this contention (e.g., Wind & Silvern, 1992; Erickson *et al.*, 1989; Kinard, 1982). In addition, some researchers have suggested that greater the number of sub-type of maltreatment (e.g., physical, sexual abuse and neglect), the more negative outcome will be for the child as well as the adult with a history of PCA (Kurtz, Gaudin, Howing & Wodarski, 1993; Wind & Silvern, 1992). Browne and Finkelhor (1986) summarized eleven research findings and reported that out of eleven, six confirmed the relationship between abuse and traumatic harms. Beitchman *et al.* (1992) suggested that abuse of longer duration is linked with long term lasting harm. Browne and Finkelhor (1986) refer to ten studies that considered both long and short-term effects of abuse, of different degrees of severity. Out of these studies, six reported linkages between harmful outcomes and the severity of the abuse however, four found no significant link. They summarized that it is premature to conclude that molestation involving more intimate contact is necessarily more traumatic than less intimate contact. Beitchman *et al.* (1992) by quoting Browne and Finkelhor studies and looking at some additional studies, concluded that there is a linkage between serious abuse and trauma as perceived by the victim, but a less strong association between such abuse and mental health was reported.

2. *Stress and Psychopathology*

Researchers suggested that negative effects of abuse are greatest for children in families in which high levels of stress and parental psychopathology, e.g. Schizophrenia or depression, are found (Kurtz, *et al.*, 1993; Walker, Downey & Bergman, 1989). Reports also demonstrated the influence of socio-cultural and family variables (e.g., SES and quality of the parent-child interaction) on negative outcome (Herrenkohl, Herrenkohl, Rupert, Egolf & Lutz, 1995).

3. *Age at Onset of Abuse*

The age, at which the onset of abuse has been referred, has also impact on outcomes. It is generally believed that child sexual abuse is less harmful in younger children, because of their lack of awareness of the social stigma attached to sexual abuse (Gomes-Schwartz *et al.* 1990). Adams-Tucker (1981) reported that children first abused after the age of 10 experienced worse short-term symptoms than those first abused before this age. By contrast, Russell (1986) found that long term ill effects were more likely to be experienced by those first abused in pre-puberty rather than in adolescence. Beitchman *et al.* (1991) suggest that age need to be considered in association with these variables.

4. *Gender of Child*

Some studies denoted the variation in the extent of abuse and neglect between boys and girls. Though, abuse of girls and boys can take different forms and this may account for different outcomes rather than the fact that the victims are boys or girls. Studies (Pierce & Pierce, 1985; Finkelhor, 1995) indicated that boys are more likely to be subjected to violence when being sexually abused and therefore, more likely to suffer negative effects. On the other hand, girls are more likely to be abused by their own family members and neighbours.

5. *Abuser-Abused Relationship*

Researchers suggest that if abuser is more closely associated to the victim, the greater degree of harm was found. Thus, intra-familial abuse is likely to have a more harmful effect than exo-familial abuse and abuse perpetrated by own father more than that of committed by other family members. Contrary to this, Browne and Finkelhor (1986) quoted several studies and compared between intra-familial abuse and exo-familial abuse, they concluded that the harmful effects were the same; however, abuse by father figures compared with that by all other perpetrators was considered to result in greater trauma (Beitchman *et al.*, 1992). They also stressed that such abuse is likely to be of longer duration and to reflect and create considerable family dysfunction.

6. *Abuse with Violence*

Investigators (Browne & Finkelhor, 1986) point out the use of violence or the threat as one of the important factors contributing to both short and long-term distress and harm. Russell (1986) reported association between these two variables. All those who reported violent abuse experienced extreme or considerable trauma, as did 74 per cent of those who experienced forceful abuse than those who experienced non-forceful abuse (47%) experienced lasting harm.

7. *Response to Abuse*

Studies report that a supportive response from the family was an important factor in reducing the extent of long term problems following sexual abuse (Conte & Schuerman, 1987). Beitchman *et al.* (1991, 1992) refer to several other studies that confirm the influence of family response on both short and long term harms. They view that victims of child sexual abuse are more likely, than non-victims, to come from disturbed families.

8. The Role of Protective Factors

Recent studies have implicated the protective influence of certain factors such as high intellectual functioning (Herrenkohal *et al.*, 1991) and the presence of a supportive parent figure (Herrenkohl *et al.,* 1995) in minimizing the consequences of abuse and neglect. Several variables have been associated in the literature with influence on the harmful effect of child sexual abuse.

Psychological Consequences of Child Abuse and Neglect

A cursory glance at examining the available literature on child abuse denotes the short-term and long-term ill consequences of child maltreatment on overall development of children, yet some psychological processes need in-depth investigation in relation to child abuse.

Thus, in present study, researcher endeavoured to explore the impact of child abuse on Cognitive, Motivational and Behavioural functioning and Personality development of children.

[1] Cognitive Process

This section focuses on the consequences of child abuse and neglect on cognitive functioning of children. Since cognitive processes are the extensively studied area of psychological researches and over the last few decades, cognitive processes have represented one of the important topics of empirical investigations in developed as well as in developing countries. Since 1971, Indian researchers have contributed a lot in the area of cognitive processes; thus, this area has achieved prestigious focus/status in the realm of psychological investigation (Mishra, R.C., 1992; Tripathi, L.B., 1988; Gupta, G.C., 1980). Now this area has rich theoretical and empirical foundation (Mishra, R.C., 1992). Researchers have made effort to sketch an account of prominent cognitive processes, citing other Indian studies, in a very systematic manner (Mishra, R.C., 1992; Tripathi, L.B., 1988).

Psychologists endeavoured to define terms "cognition" and "cognitive processes". Cognition refers to any and all forms of various modes of knowing and understanding the world around us. It contrasts with the affective, volitional, viz. feeling and 'will' aspects of conscious life. It is a mediational process. According to Neisser (1967), "cognitive process refers to all the processes by which the sensory input is transformed, reduced, elaborated, stored, recovered and used". It includes a cluster of processes through which we get knowledge about our environment and apply this knowledge to the solution of day-to-day problems.

The continuum of cognitive processes ranges from the simple perceptual to more complex thinking and reasoning processes. In other words, sensation, perception imagery, retention, recall, problem solving and thinking are the various aspects of cognition. Apart from this, recognition, labelling, analysis, planning and categorization are also some of the cognitive processes and viewed as intellective processes through which individuals try to comprehend their environment and solve numerous problems that they encounter with environment. Primarily, the topic of cognition has its historical root in the work of Freud, Toleman, Wertheimer, Lashley and Bartlett. Later on Lewin, Piaget, Weber, Bruner and Witkin contributed in the field of cognitive development. It seems that cognition has very strong theoretical base and has proved its significance in adaptation of people with their social and cultural environment.

Since researches on cognitive processes are concerned with two broader traditions. First is the tradition of hard-core experimental researches in which emphasis is given on describing how an individual receives information, transforms into meaningful codes, recollects whenever needed and, apply them to solve the problems. Whereas, second tradition includes developmental researches, in which researchers try to know how growth and development of cognitive processes occur from maturational and socio-cultural perspectives. In fact cognitive skills develop gradually and much of shaping depends upon environment hence there is reciprocal relation between cognition and environment.

Studies denoted that cognitive development proceeds through clearly defined stages in predictable manner. However, cultural transmissions can accelerate or slow down the development of cognitive functions (Piaget, 1972). Therefore, some variables are found during pre-operational, concrete operational and formal operational stages in different socio-cultural context. In a study the effect of home environment and caste on development of conversation ability of primary school children was examined. Results revealed that Brahmin children performed superior than Harizans. Further, children from poor home environment were found inferior than those belonging to enriched environment (Behera, 1989). In another study (Gupta, 1991) deprived and non-deprived adolescents of high and low SES students were identified. The finding revealed differences in their personality disposition, adjustment level, intelligence and academic achievements between adolescents of deprived home environment and non-deprived counterparts. Niraula and Mishra, R.C. (2001) compared different age levels of boys and girls belonging to rural and urban communities on perceptual differentiation task (EFT). Urban children scored higher than those of rural children; however gender difference was non-significant (Niraula & Mishra, 2001).

Perceptual Identification

In a study, Verma, Upadhyay and Mishra (2003) examined the effects of task structure on the development of perceptual exploration patterns in children between 3-12 years of age. Results revealed that errors of omission and commission decreased and with increasing age, the number of children identifying correctly, increased. They pointed out that exploration of structured stimuli is a joint function of the child's level of perceptual development and the nature of perceptual task.

Learning and Memory Process

Recent studies in the field of memory pointed out some strategies related to reception, acquisition, coding, organization and retrieval of information in determining retention and recall (Bjorklund, 1988, Ackerman, 1985). Psychologists have investigated the roles of certain socio-demographic factors in memory process. Donga (1989) investigated the effect of certain personal and social factors such as age, sex, grade, socio-economic status and family size and a set of memory task/tasks were given to students of class III and class XII. Results revealed that age, birth order, family size and SES exercised impact on memory, however, the impact of grade level and gender was non-significant. Verma (1988) studied the impact of motivation and personality along with age, sex, and sense modality on S-T-M and L-T-M in children of 12-20 years age. Results revealed better short term and long term memory in older than younger age group and visual than auditory presentation condition, high motivation group showed better memory than low motivation group. Extroversion/ Expression was found to be positively correlated with STM of 12 to 16 years of age children. Researchers denoted that variability in memory is found due to strategies used in acquisition and processing of information. Dixit (1988) studied a set of 35 strategies of information acquisition in relation to various socio-demographic and personality factors, and pointed out 11 factors out of 35, to account for variance in performance. These were: Deep processing, elaborative processing, fact retention, information dependence and prediction orientation, knowledge of results, methodological study and inferential measurement. The type of family, parental income and education were found significantly associated with the selection of strategies.

Creativity, Reasoning and Intelligence

Researches denote that cognitive process develops from undifferentiated to a differentiated level. Witkin *et al.* (1962) pointed out that heterogeneity of functions is the most important characteristic of differentiation, which is reflected in the segregation of psychological functions. In early stages basic structure helps to accomplish variety of functions whereas complex abilities like reasoning includes a variety of structures, which may be organized informally, or differentially across group depending on whether group represents culturally homogenous or heterogeneous units (Irvine & Berry, 1988). Since, cultural diversity is one of the common characteristics shared by societies worldwide. The structure of abilities is supposed to vary across groups. This assumption was studied by Suri (1989), in assessing the reasoning ability of rural and urban students. Findings revealed that cognitions of semantic classes, semantic relation and convergent production of semantic inflictions are three factors accounting for the reasoning ability of rural children, whereas, convergent production of semantic classes emerged as the only factor to account for students reasoning ability. In another study, Garg (1988) analyzed the factorial structure of the reasoning abilities of 14+ years of male students. 37 tests of reasoning abilities were administered. Factor analysis results revealed various factor structures for heterogeneous and homogeneous groups of structure. Results further indicated that differentiation ability occurred around age 14+ and it was possible to identify differentiated dimension of reasoning abilities. In a factor analytic study of 25 structure of intellect including two tests of creative thinking yielded five prime factors, which accounted for a major portion of variance in the intellectual performance of high school students.

Dash (1991) in his study emphasized on the processes of good and poor readers with high and low I.Q. children of class III. Results revealed that good readers were superior to poor readers in selective attention, figure memory, serial recall and verbal processing factors requiring successive or simultaneous processing. However, no significant differences were found on speech related task and I.Q. of good and poor readers.

In a study (Dei, 1991) the relationship between non-verbal measures of cognition and gender differences in performance was examined on Wechsler's Digit Span Test, Ravens colour progressive metrics and Non-language test of working memory for class Vth boys and girls of low socio-economic status. The results indicated non-significant differences between boys and girls on non-verbal cognitive measures, however, the interrelationship among above said three measures were positive and significant.

Intellectual ability and commitment jointly exercise impact on educational interest of students. Kumari (1986) investigated the relationship of intellectual commitment and educational interest with intelligence, SES, academic achievement and some personality traits. Results revealed that higher intellectual commitment was related to father's low education, whereas, lower intellectual commitment was correlated with father's higher education, high intelligence and SES. Intelligence and introversion were found positively correlated in boys and on the other side, intelligence and intellectual commitment were positively related with academic achievement of girls.

Cognitive Functioning in Relation to Child Abuse and Neglect

A sizcable numbers of researchers have established the impact of various forms of cognitive processes on child abuse and neglect.

(i) Impact on Cognitive/Intellectual/Academic Functioning

Several studies have documented that physically abused children (PAC) show lower intellectual and cognitive functioning relative to a comparison group of children on general intellectual functioning as well as specific measures of verbal facility, memory, verbal language, communication abilities, problem solving, and perceptual motor skills (Fantuzzo, 1990; Friedrich, Einbender & Luecke, 1983; Haskett, 1990; Hoffman-Plotkin & Twentyman, 1984). Academic performance is another area of substantiated difficulty in physically abused children. It is evident from a sizeable number of studies that physically abused children as compared with non-abused children, display poor school achievement and adjustment, receive more special education services, score lower on reading and math tests, exhibit more learning disabilities, and are more likely to repeat a grade (De Paul & Arruabarrena, 1995).

(ii) Intellectual/Academic Deficiencies in Abused and Neglected Children

A large number of studies comparing neglected children and adolescents to non-neglected groups indicated that neglected victims show deficits in language abilities, academic skills, intelligence and problem-solving skills (Egeland *et al.*, 1983; Hoffman-Plotkin & Twentyman, 1984; Wodarski *et al.*, 1990). Child maltreatment damages the school performance of children. Wodarski *et al.* (1990) evaluated 139 school-age adolescents physically abused, neglected and non-maltreated children and found that both neglected and abused children were significantly poorer on overall school performance and math skills. However, the neglected children expressed poor language and reading skills than the group of non-maltreated children. These intellectual deficits continued even after the researchers controlled the influence of socio-economic disadvantages.

Exposure to severe child physical abuse may produce cognitive or intellectual deficits or other related problems in academic achievement. Certain early studies showed reduced task intimation and motivation (Aber & Allen, 1987; Allen & Tranowski, 1989), limited intellectual functioning among maltreated youngsters (Alessandri, 1991; Erickson *et al.*, 1989) and school children (Salzinger *et al.,* 1984) whereas, other studies have failed to find differences in these areas (Augoustions, 1987; Azar, 1997; Fantuzzo, 1990; Slade, *et al.* 1984). Language deficits have also been found in both receptive and expressive areas, including more limited syntactic expression, functional communication, and self-related language (Coster *et al.*, 1989; McFayden & Kitson, 1996).

Problems with academic performance also include low scores on school discipline, referral and suspensions (Eckenrode, Laird & Doris, 1993). Overall, maltreated children were 2.5 times more likely to repeat a grade. These effects were not attributable to socio-economic status (SES), gender, or age. As suggested by researchers, the academic careers of maltreated children may be marked by considerable discontinuity due to frequent moves, school transfers and tardiness. In contrast, no relation between child physical abuse and school grades or receptive language was found in a large community study (Fisher *et al.*, 1997).

The varied outcomes in this area may be attributable to several background factors such as the failure to distinguish among types (severity) of maltreatment or to use objective measures (Augoustions, 1987). Cognitive functioning may be closely related to the quality of the care taking environment, as shown in one study that found that young victims of physical abuse had more receptive language problems than non-abused children but did not differ from children residing in low SES families (Vondra, Barnett & Cicchetti, 1990). Given that the degree of enrichment in the home environment influences a child's cognitive abilities (Aber *et al.*,1989), the impact of family environmental variables may be considerable. The impact of physical injuries on children's cognitive abilities and their level of academic achievement merits further evaluation.

(iii) Perceptual and Attributional Problems

Certain perceptions or attributions related to the experience of physical violence have been reported among children with a history of child physical abuse. Relative to controls, abused children have reported a greater willingness to use physical punishment (Carlson, 1986, 1991), were more likely to assimilate aggressive stimuli (Reider & Cicchetti, 1989).

Although these few studies provide some understanding of the cognitive deficits or distortion evinced by abused children, there is a need to more carefully examine the cognitive repertoires of physically abused children. Potential cognitive problems include their belief in the appropriateness of violent behaviour as a coping skill and their difficulties identifying non-aggressive solutions to interpersonal problems. It would be instructive to learn, for example, whether victims to child physical abuse show negative attribution biases comparable to those of aggressive children (Dodge, Bates & Pettit, 1990) or the types of cognitive distortions commonly seen in depressed children (Kendall, Stark & Adam, 1990). If so, then specific method designed to enhance children's use of pro-social, verbal mediation skills would be likely to affect these cognitive processes.

(iv) Cognitive Difficulties and Distortions in Sexually Abused Children

Cognitive functioning may be affected by sexual abuse experiences. Einbender and Friedrich (1989) reported greater cognitive impairment in sexually abused girls whereas Rust and Troupe (1991) reported sexually abused children have deficits in school achievement. Negative or distorted attribution and perception have also been frequently observed (Hunter *et al.*, 1992).

Sexually abused children exhibit numerous cognitive problems: like (1) Academic problem, (2) Intellectual delay, (3) Lower level of overall intelligence, (4) Receptive and excessive language deficits, (5) Less creative and flexible in problem solving, and (6) Deficits in language comprehension/and verbal abilities, etc.

Table 1.3 : The consequences of sexual abuse vary across different age groups

Pre-school	*School Age*	*Adolescent*
Learning Difficulties	Learning difficulties*, poor concentration, Poor attention, Declining grades, Negative perceptions, Dissociation	Learning difficulties, Poor concentration, Poor attention, Declining grades

* Indicates most common symptoms for age group.

Table 1.3 shows possible short-term cognitive effects associated with sexual abuse in pre-school, school age and adolescent children suggested by numerous studies (Boney Mc-Coy & Finkelhor, 1995; De Bellis, Burke, Trickett & Putnam 1996; De Bellis, Chrousos *et al.*, 1994).

(v) Cognitive Deficiencies in Psychologically Abused Children

Child characteristics likely to interfere with cognitive function have been found to be associated with psychological maltreatment in many of the studies. The conditions identified include: decline in mental competence, lower intelligence, non-compliance, lack of impulse control, impaired learning, academic problems and lower achievement level and impaired development of moral reasoning (Claussen & Crittenden 1991; Crittenden *et al.*, 1994; Egeland & Erickson, 1987; Hughes & Graham-Bermann, 1998; Main & Goldwyn, 1984; Main & George, 1985; Rohner & Rohner, 1980).

Several other studies have found that rejected children perform less well than comparison children, on achievement and intelligence tests, perform poor in school (Chan, 1981; Manley, 1977; Sheintuch & Lewin, 1981; Strakey, 1980) and exhibited impaired development of moral reasoning (Bilbro, Boni, Johnson & Roe, 1979; Hoffman, 1970). Hart and Brassard (1993) found that psychologically maltreated children from a disadvantaged group were more likely to show school-related problems in ability and academic achievement.

[2] Motivational Process

Consequences of Child Maltreatment on Motivational Process

Present researcher has made effort to investigate the role of child abuse and neglect in achievement motivation and level of aspiration in children.

Motivation refers to a process which initiates, directs and maintains progress towards the goal. A goal is a mental representation of something we need (National Advisory M.H. Council, 1995). It can be general or specific. Psychologists, trying to understand the motivation behind behaviour; focus their researches in one or two ways. From an individualist perspective, they study the person on independent entity. Whereas, from the collectivists perspective, they view the person as inseparable from his or her relationship with other and both perspectives have different ways of conceptualizing a person (Triandis, 1995). Mostly, psychological studies on motivation, based on individualistic perspective, have tried to identify the personal need and thoughts that produce goal oriented behaviour. Motive is classified into two categories : (i) Biological motivation and, (ii) Psychological motivation. Abraham Maslow arranged five basic needs into the "hierarchy of need" in accordance with their relative importance for survival. These needs include, (i) Physiological needs, (ii) Safety and security, (iii) Belongingness, love and affection, (iv) Self-esteem and (v) Self-actualization. The Biological conceptualization of motivation come first and behaviours as eating, protecting, aggression etc., have biological roots whereas psychological motives are learned or acquired and therefore are known psycho-sociogenic motives. These motives are mainly acquired through experiences of socio-interaction at various levels and once learned, the motives influence the way in which person evaluate the demands and challenges raised by social motive, since the development of several psycho-sociogenic motives are largely determined by opportunities and experiences available to the person, it is not unrealistic to believe that individuals having varying levels of experiences would show differences in the motivational patterns. It was therefore, planned to investigate the level of achievement and aspiration in relation to child maltreatment.

(i) Achievement Motivation

One of the prominent sociogenic motives is achievement motivation. This is the desire to excel and accomplish a significant goal. It is caused by need for achievement. A great deal of empirical investigations has been done in the field of achievement motivation (Atkinson, 1964; McClelland *et al.*, 1953, Feather, 1961). The experimental study of achievement motivation has its origins in Murray's taxonomy of need systems and his development of a projective test, the TAT, to infer an individual's motivational concerns. The TAT was refined by McClelland and his colleagues for use, as a research tool to assess the strength of achievement needs. This tool is extensively used by researchers (Agarwal & Tripathi, 1980).

The needs for achievement were then included as one component of a theory of achievement motivation. This theory was formulated by Atkinson and specifies that achievement-related behaviour is result of conflict between hope of success (approach motivation) and fear of failure (avoidance motivation). The approach and avoidance tendencies, in turn, are functions of achievement-related needs (need for achievement and anxiety about failure), the expectancy of success and failure, and the incentive value of success and failure (TS = MS×PS×IS *vs.* Taf = Maf×Paf×Iaf). A special assumption of the model is that the incentive value of achievement tasks is determined by the probability of success. The main derivation of this theory is that individuals high in resultant achievement needs are particularly attracted to tasks of intermediate difficulty, while those low in achievement needs avoid tasks that have a probability of success near 50 per cent there is suggestive evidence that tasks of intermediate difficulty are more attractive to individuals who are highly motivated to succeed than to those lower in achievement needs. In a similar view, individuals high in achievement need have been characterized as "realistic" and have occupational goals that are congruent with their abilities (Mahone, 1960; Morris, 1966). The desire for intermediate risk may be indicative of a preference for personal feedback or knowledge about oneself. This informational explanation is consistent with the high achiever's reported preference for business, occupation where feedback (projects) is evident (Mayer, Walker & Litwin, 1961; McClelland, 1961). In addition, individuals' high in need for achievement apparently are better able to delay gratification (Mischel, 1961) and attain higher grades in school than individuals low in achievement needs, if the grades are instrumental to long-term success (Raynor, 1970). Furthermore, individuals high in achievement needs are conceptualized as "hope" rather than "fear" oriented. High achievers are found future oriented than low achievers (Agrawal & Tripathi, 1980). Finally, individuals high in achievement needs take personal responsibility for success and generally; perceive themselves as high in ability (Kukla, 1972). The self-attribution for success increases their feeling of worth; they volitionally undertake achievement-oriented activities when the opportunity arises (Atkinson, 1953; Green, 1963). In addition, the self-perception of high ability may account in part, for their high self-concept. (Mukherjee & Sinha, 1970).

Achievement behaviour is rewarding partly because it is productive. People who achieve significant goal feel good about themselves and their efforts. In some situations they also improve the lives of other. The particular goal that people aim for, are shaped by their experiences. For some students, academic achievement bolsters self esteem, they feel pleased with themselves when they achieve high grades and feel unhappy with themselves when they fail to get. Some students however, loose their reactions to academic performance, partly because of their previous experiences. Osborne (1997) found that African-American boys were more likely than the others to become detached from their academic achievement (Osborne, 1997).

Impact on Achievement Motivation

Empirical studies indicated negative linkages between child maltreatment and achievement levels of children. Brassard, Hart and Hardy (1991) found that psychologically maltreated children from disadvantaged group expressed school related problems in ability and academic achievement. Similarly, neglected children performed less well than comparison group on achievement test (Sheintuch & Lewin, 1981, Strakey, 1980). Thus, in present study it was considered pertinent to investigate the impact of child abuse on achievement level of children.

Level of Aspiration

Kurt Lewin developed a variant of expectancy value theory, known as resultant valence theory, within the context of the study of level of aspiration. Level of aspiration typically is defined as "the level of future performance in a familiar task which an individual knowing his level of post performance in the task, explicitly undertakes to reach" (Frank, 1935). Level of aspiration thus, pertains to goal striving and perceived difficulty of the goals that one wishes to attain.

Numerous empirical findings have emerged from level of aspiration research. It has been demonstrated that feeling of success and failure are determined primarily by the attainment or non-attainment of the goal for which one is striving, as opposed to the absolute level of performance. For example, a student with a very high aspiration level might be unhappy upon receiving grade B while an equally bright student with lesser aspiration wound be satisfied with this grade. Perception of success and failure involves subjective, rather than objective levels of attainment. William James stated that feelings of success or self-esteem are a direct function of actual success and an inverse function of pretensions. Further, the subsequent aspiration level is, in part dependent upon the prior attainment discrepancy (subjective success or failure). In most instances, the goal level increases after success and decreases following failure. However, "atypical" reactions are sometimes observed; that is, sometimes an individual lowers his or her aspiration level after goal attainment or raises it after failure.

Individual differences, group standards, and cultural factors influence level of aspiration. For example, individual with high ability, or those classified as ambitious, tend to set higher aspiration levels (Hoppe, 1930). In addition, the level that is set tends to be guided by, and conforms to group goals and performance (Festinger, 1942). Mishra and Tripathi (1980) found negative association between prolonged deprivation and level of aspiration. Further, deprived subjects expressed lower level of aspiration than their low deprived counterparts.

The concept of level of aspiration has also been modified and used for the analysis of social interaction and interdependence. Thibaunt and Kelley (1959) assumed that individuals can assess their degree of satisfaction with a relationship, just as one can determine the valence of an achievement related success or failure. They introduced the concept of "comparison level" (CL) to define "a neutral point on a scale of satisfaction-dissatisfaction". If the average of "outcomes from a relationship" is above the comparison level (CL), the relationship is "rewarding"; however, if the outcome average is below the CL, the relationship is relatively punishing. However, even if the relationship is negative, individuals may remain interdependent. The continuation depends, in part, on the alternatives that are available or the potential satisfaction from equivalent relationship or opportunities. In sum, Thibaunt and Kelley assume a rational person striving to maximize gains, capable of comparing a number of potential satisfiers and deciding which "goal" (relationship) to pursue. This is clearly a direct extension of aspiration work into the sphere of interpersonal relations.

Therefore, it was assumed that family environment, pattern of parent-child relationship and abusive maltreatment by parents or caregiver, would exercise negative role in development of level of aspiration in children.

[3] Behavioural Process

The impact of child abuse on behavioural functioning of children is widely investigated area Several behavioural indicators of child physical abuse (CPA), child sexual abuse (CSA) and child psychological abuse (C Psy A) and child neglect have been identified: In present study aggression and helping behaviour were studied in relation to child abuse.

Aggression Level in Children

Often aggression is viewed as any behaviour that hurts other. The advantage of this is that the behaviour itself determines whether or not an act is aggressive. Ignoring intention can also produce the opposite error calling some acts aggressive that are not by the usual meaning of the term. Many acts that are painful are intended to help the victim, such as surgery performed by physicians. Thus, aggression is defined as, "any action that is intended to hurt other." The second major distinction is needed between antisocial and pro-social aggression. Some aggressive acts fall somewhere between pro-social and antisocial and we might level them sanctioned aggression. This includes aggressive acts that are well within their boundaries. They do not violate accepted moral standards. A third useful distinction is between aggressive behaviour and aggressive feelings such as anger. Our overt behaviour dose not always reflects our internal feelings. Society discourages and condemns most forms of aggressive behaviour and indeed can exist only if people control their aggressive feelings most of the time. An aggressive feeling is an internal state that cannot be observed directly. People experience anger and virtually everyone at one time or another would like to hurt some one else. Indeed, most people report they feel at least mildly or moderately angry anywhere from several times a day to several times a week (Averill, 1983).

Two main factors concerned with aggression are: Attack and Frustration. One of the most common sources of anger is being attacked or bothered by another person. It is depending on how the injured person takes it; he or she has been annoyed or attacked. It is likely that the person will become angry and feel aggressive toward the source of the attack. A second major source of anger is frustration. Frustration is the interference with or blocking of the attainment of a goal. One of the basic tenets in psychology is that frustration tends to arouse aggressive feelings. Researchers asserted that aggression is always a consequence of frustration, the occurrence of aggressive behaviour always presupposes the existence of frustration and the extents of frustration always lead to some form of aggression (Dollard *et al.*, 1939).

Family life is one of the major sources of frustration, surprisingly; the most commonly reported source of conflict. Economic problems produce especially high level of frustration within families. There is a more family conflict and more domestic violence in working class than in middle class families, and more in families with unemployed breadwinners or with especially large number of children (Straus *et al.*, 1980). Job related problems are also the greatest sources of frustration and anger.

These examples illustrate the typical effect of frustration. But original theory suggested that aggression always stemmed from frustration, and frustration always produced aggression (Dollard, *et al.,* 1939).

In most cases, for an attack or frustration to produce anger, aggressive behaviour must be perceived as intended to harm. In terms of causal attribution, anger is most likely when the attack or frustration is perceived as intended by the other person that is as being under that person's internal control (Weiner, 1982). Therefore, several theories have tried to explain to the origin of aggressive behaviour. A sizeable number of studies indicated that a cluster of personal, social and situational factors causes aggression.

In the following section an effort has been made to present a brief account of researches on the consequences of child abuse and neglect in the forms of heightened aggression and other externalizing problems in children.

Heightened Aggression in Abused Children

One of the most extensively documented consequences in child physical abuse victims is heightened aggression and related externalizing behaviours (National Academy of Sciences, 1993), including poor anger modulation (Beeghly & Cicchetti, 1994; Shields, Cicchetti & Ryan, 1994). Group differences have been reflected on adult ratings (Cummings, Hennessy, Radideau & Cicchetti, 1994; Dodge, Pettit & Bates, 1997; Feldman *et al.*, 1995; Hotaling, Straus, Lincolin, 1990; Manly, Cicchetti & Barnett, 1994; Okun, Parker & Levendosky, 1994), self-ratings (Pelcovitz, Kaplan, Goldenberg & Mandel, 1994; Wolfe, Werkele, Reitzel-Jaffe & Lefebvre, 1998) and observations in both the home (Bousha & Twentyman, 1984) and play or social setting (Alessandri, 1991; Howes & Espinosa, 1985; Kaufman & Cicchetti, 1989; Klimes-Dougan & Cicchetti,1989). Dodge *et al.* (1997) pointed out that the experience of child physical abuse included aggressive behaviour and conveys its appropriateness as a problem-solving skill.

The level of externalizing dysfunction may be related to the experience of witnessing family violence and other adverse family environment factors, such as negative life events (Wolfe, 1999). Both heightened family conflict that includes verbal and physical aggression (*vs.* verbal only) and shelter residence (*vs.* home) have been found to be significantly related to the clinical severity of children's conduct and emotional problems and diminished social functioning (Fantuzzo *et al.*, 1991). Pandey (2004) investigated the influence of child abuse and neglect on aggression level in children. Results revealed that child abuse was found to be positively correlated with aggression. Aggression was found to be high in physically abused children particularly belonging to low socio-economic status. Physical aggression and antisocial behaviour are among the most common correlates of child physical abuse (CPA) (Wolfe, 1987). In most studies, abused children show more aggression, even after the poverty, family instability, and wife battering that often accompany abuse is statistically controlled (Fantuzzo, 1990).

Apart from this, certain other indicators of physical and sexual abuses are identified. These outcomes are:

(i) *Aggression and Acting out:* Running away from home committing criminal acts or turning to prostitution are behaviours common to abused children. The majority of runaways are found from neglectful, abusive and violent families. Children who become involved in prostitution are overwhelmingly characterized as having a history of sexual abuse. Their low self-image combined with a belief that sex is their only valuable asset, makes them natural pray for exploitation through prostitution. A child's entry into delinquency is often a product of a dysfunctional family life and subsequent susceptibility to negative peer pressure. The violence or severity of criminal acts by young offenders seems to be correlated with the level of violence and neglect they experienced at home. Children who have been physically abused are prone to acts of aggression against younger children and animal, defiance against authority figures and exaggerated temper outburst.

(ii) *Regression or Pseudo-maturity:* The stress from physical or sexual abuse together with demands that exceed the child development capacity can result in regression. Children acting younger than their chronological age characterize regression under pressure to act older than their age; some victims develop a preconscious adult-like behaviour.

(iii) *Grooming Habits Hygiene:* Child's grooming habits and hygiene may sign body shame or fear of detection of the abuse.

These changes may include dressing with clothing that hides the body, especially when it is out of season, refusal to bath or bathing excessively, or an abrupt increase in modesty. Sexually abused children may have the mistaken belief that their bodies are somehow marked by their sexual experiences, while physically abused children may seek to hide obvious bruises and belts. Some sexual abuse victims become obsessive about their hygiene.

(iv) *Eating Problem:* Studies report that some sexual abuse victims may develop eating disorders, such as anorexia (starving oneself) and bulimia (bingeing and purging). A child's wish to change his or her appearance to avoid being attractive to the perpetrator may result in the child refusing to eat. Neglected children may gorge or hoard food when it is available because they are uncertain when they will be fed next. Some psychologically maltreated youngsters are told they are fat or ugly and develop eating problem in response to these negative messages. Other abused children find food intake is the only aspect of their lives that they can control.

(v) *Sexual Behaviour:* Perhaps the strongest behavioural indicator of sexual abuse is developing inappropriate sexual behaviour and sexual knowledge. These children may be sexually preoccupied, make drawings and tell stories with sexual content or tell sexual jokes. Their conversations may contain frequent or unusual sexual references and their activity may be focused around their sexualized interests. It is common for children to be curious about body parts, have a desire to look at another's body to ask questions about procreation. However, research shows that there are a number of sexual behaviours that occur with such low frequency that they can be considered red flag sexual abuse. Some sexually abused children reenact their experiences with siblings, pets and peers. If a child persists in sexual behaviour after being told that behaviour is essential to protect the child.

(vi) *Post-Traumatic Responses:* When physical or sexual abuse overwhelms the child's coping ability, leaving the child feeling helpless terrified and over stimulated, traumatic responses may follow. A child's response to events that appear traumatic is based upon the child's resilience and ability to cope during the trauma. The three most common post-traumatic reactions involve experiencing; avoidance and increased physiological arousal; re-experiencing involves repetition of the trauma in dreams, thoughts play or action. Flashbacks and reenactments of the abuse are also forms of experiencing. An avoidance reaction includes efforts not to be reminded of the abuse by avoiding thoughts, place and people.

(vii) *Hostility and Aggression in Sexually Abused Children:* Sexually abused children respond by directing anger and aggression outwards. The response is more common among adolescents. Gomes-Schwartz *et al.* (1990) identified such a response in between 45 per cent to 50 per cent of the 7 to 13 years old in their sample and Friedrich *et al.* (1986) found that children aged from 6 to 12 years more likely to externalize their feelings than younger children. They point out that there are problems associating aggressive reactions purely with sexual abuse because many of the children in their sample were also subjected to physical violence or the threat of it. Thus, the aggression they displayed could have been a result of this rather than of the sexual abuse they experienced.

(viii) *Self-injurious Behaviour:* Some abused children demonstrate self-inflicted acts such as burning, cutting, disfiguring or fair pulling. These behaviours may reflect a desire to punish them for not stopping the abuse or an attempt to gain some control over their own bodies or to feel alive through the infliction of pain when they feel dead inside.

(2) Helping Behavior

Helping behaviour is a popular form of pro-social behaviour. Altruism and helping behaviour are two key concepts, which are highly rated in our society, however differ on some aspects:

(i) *Altruism :* Refers to an act performed voluntarily to help some one else when there is no expectation of receiving a reward in any form (except perhaps a feeling of having done a good deed). By this definition, whether or not an act is altruistic depends on the intentions of the helper.

(ii) *Helping Behaviour :* Helping behaviour, on the other hand, is much broader category. It includes any act that helps or is designed to help others, regardless of the helper's motives.

> Many pro-social acts are not altruistic. Our everyday experiences provide many examples of pro-social behaviour. Even young children show the capacity to be helpful to others (Zahn-Waxler, *et al.*,1995).

The type of relationship between people affects pro-social behaviour. Whether because of linking social obligation, self-interest or empathy, we are more helpful to those we don't know. Helping behaviour is a widely studied area in social psychology.

The understanding of pro-social behaviour has benefited from several broad theoretical perspectives. First, the decision-making perspective focuses on the process that influences judgment about when help is needed. It also emphasizes the weighing of cost and benefit in the decision to give help. Secondly, learning approach reveals that people learn to be helpful, following basic principles of reinforcement and modelling. Third, socio-biologists have proposed that a predisposition to help is part of our genetic, evolutionary heritage.

A sizeable number of studies indicated that a cluster of personal, social and situational factors determine pro-social behaviour. It is not, necessary that in every case bystander intervention, to help the victim will be successfully implemented. Since, helping requires a decision making process, which includes multiple stages. Helping is costly in terms of time, effort, money and sometimes life; thus, interruptions at every stage of helping are very natural.

Impact on Helping Behaviour

It is quite possible that helping tendency in abused children might be damaged due to poor modelling and unfavourable circumstances in the abusive families. Pandey (2004) showed that abused children belonging to low socio-economic status expressed poor helping behaviour than their non-abused counterparts. Other studies also report diminished helping tendencies in maltreated children.

One of the most often-cited problems associated with child abuse and neglect is difficulty in social skill and adjustment. As evidence of social difficulties, a number of studies have revealed a relationship between neglect and disturbed pattern of infant caretaker attachment (Crittenden, 1992; Bousha & Twentyman, 1984; Egeland & Sourfe, 1981).

Other indication of disturbed parent-child interaction have

appeared demonstrating the deficits in communication, increased aggression and poor involvement characteristic of the interaction between neglecting mother and their children (Crittenden, 1992; Christopoulos *et al.*, 1988; Bousha & Twentyman, 1984). Additional research has also revealed the other area of social maladjustment in victims of child neglect such as socially withdrawn behaviour (Crittenden, 1992; Hoffman-Plotkin & Twentyman, 1984). They observed other social difficulties in the form of: disturbed parent-child attachment and disturbed parent-child interaction. Apart from this, abused children were found :

(1) Passive, show deficits in pro-social behaviour and communication, and display physical aggression;
(2) Parent exhibits less sensitivity, appears more withdrawn and uninvolved, uses less general speech and phrases of acceptance, uses more direct imperatives, has or displays low rates of social interaction and verbal instruction.,
(3) Disturbed peer interaction such as deficits in pro-social behaviour, social withdrawal and isolation.

On the basis of review of literature, mentioned above, it was considered pertinent to investigate the consequences of abuse on helping behaviour and aggression in children.

[4] Personality Patterns

It was formerly believed that the personality pattern was the product of heredity and that the child was a "Chip off the old block". There are ample evidences that the personality pattern is the product of both heredity and environmental influences. As Thomas *et al.* have defined, "personality is shaped by the constant interplay of temperament and environmental factors". They explain further, "If the two influences are harmonized one can expect healthy development of the child, if they are dissonant behavioural problems are utmost sure to ensure". According to Allport, "Personality is the dynamic organization within the individual of that psychological system that determines the individual's unique adjustment to the environment."

Some Psychologists define personality as the relatively stable combination of beliefs, attitudes, values, motives, and temperament and behaviour patterns arising from underlying, internal inclination that an individual exhibits in various situations. The way Sherlock Holmes analyzes, "people is part of his personality". Individual's explanations of people's behaviour place greater emphasis on person's stable internal behavioural tendencies and situational determinants (Markus & Kitayama, 1998). Studies on development of the personality pattern have revealed that three factors are mainly responsible for its development, i.e. hereditary endowment, early experiences within the family and events in later life. The pattern is closely associated with the maturation of the physical and mental characteristics, which constitute the individuals hereditary endowment. These characteristics provide the foundation on which the structure of the personality pattern is built through learning experiences.

Conceptualization of Personality

The conceptualization of personality is complex and wide yet psychologists have made effort to define the nature of personality applying various perspectives. Some of these perspectives are based on empirical researches while others on theories or clinical case studies and several others as combination of the two. These diverse viewpoints facilitate in understanding the nature of personality. Six popular perspectives of personality are: (a) Trait approach, (b) Psychodynamic, (c) Learning, (d) Social cognition, (e) Genetic and (f) Humanistic approach. Additionally, researches on influences of culture, gender, race and social power have broadened the scope of personality studies.

One way to think about personality is to compare people's personality traits, general characteristics and disposition that are presumably the basis for an individual's behavioural patterns. Raymond Cattell called that surface traits readily evident, observable personality characteristics such as assertiveness or a strong tense. Whereas, source traits are implicit and unobservable personality traits. Cattell found 16 clusters of responses, or personality dimension, that he regarded as source traits (Cattell *et al.*, 1950). These are: Reserved *vs.* Outgoing; Concrete thinking *vs.* Abstract thinking; Easily upset *vs.* Emotionally stable; Calm, submissive *vs.* Dominant; Serious, Sober *vs.* Happy-go-lucky; Enthusiastic, Rule-defying *vs.* Conscientious; Shy *vs.* Bold; Tough-minded *vs.* Sensitive; Tender-minded, trusting *vs.* Suspicious; Practical *vs.* Imaginative; Forthright *vs.* shrewd; Worldly Self-assured *vs.* Apprehensive; Insecure, conservative *vs.* Experimenting; Group dependent *vs.* Self-sufficient; Undisciplined *vs.* Self-control and Relaxed *vs.* Tense. People vary in the degree to which they have these traits. For example, every person could be described by her or his position along with the dimension, defined by reserved at one end and outgoing at the other, although some people will be more reserved than other people.

Prior to R.B. Cattell, British psychologists Eysenck used factor analysis to identify personality dimension based on own personality questionnaires. Eysenck described two dimensions; introversion-extroversion and stability-instability of personality (Eysenck & Eysenck, 1963).

Recently, psychologists have developed a personality measure known, The "BIG FIVE". Numerous psychological studies have produced five traits/dimensions or clusters (Costa & McCrae, 1997; Heaven, 1996; McCrae & Costa, 1990). Therefore, they called Big Five dimensions of personality, which are found across cultures and both sexes (Paunonen *et al.*, 1992) are as follows : (i) Extroversion, (ii) Conscientiousness, (iii) Openness to Experience, (iv) Emotional Stability, and (v) Agreeableness.

However, everyone's personality cannot be adequately elaborated within five dimensions. Still Cattell's personality questionnaires are popularly used to study the personality structure of people.

Psychologists have explored a set of factors, viz. intelligence, physique, attractiveness, childhood experiences, feelings, emotions, motive of success and failure, cultural influences and family influences etc., exercising impact on personality development. Researchers have extensively studied the roles of family dysfunctions, inadequate treatment by parents or caregivers and poor bond between parent-child relationships as predictors of child maltreatment, which in turn damage the personality development of children.

Consequences of Child Abuse and Neglect on Personality Development

Child physical abuse and personality disorders are found closely associated with several problems, viz. major depression, conduct disorder, oppositional defiant disorder, agrophobia, over-anxious disorder, and generalized anxiety disorder (Fisher *et al.*, 1997). Higher rates of depression and conduct disorder have been found in abused adolescents (Pelcovitz *et al.*, 1994). A history of maltreatment has also been associated with borderline personality disorder (Famularo, Kinscherff & Fenton, 1991), attention deficit, hyperactivity disorder and appositional disorder (Famularo *et al.*, 1992). This latter study is noteworthy for its finding that maltreated cases differed from controls in certain disorders depending on whether children (e.g., greater psychotic symptoms, personality/adjustment disorders) or parents (conduct/mood disorders) were interviewed. Physical abuse has been found to contribute to the prediction of current diagnosis of several disorders, cigarette smoking, as well as other lifetime disorders and problems (e.g., conduct disorder, drug abuse), when added to other identified risk factors (Kaplan *et al.*, 1998). Finally, traumatic victimization (e.g., assault, mugging and sexual molestation) has also been found to contribute to the prediction of oppositional defiant disorder but not attention deficit/hyperactivity disorder in outpatient child psychiatry clinic referrals (Ford *et al.*, 1999).

Child Sexual Abuse and Personality Disorders

Researchers have identified numerous short-term and long-term consequences of child sexual abuse.

(i) *Low Self-Esteem, Guilt and Shame :* The findings with regard to low self-esteem as a short-term consequence of sexual abuse are rather equivocal. De Francis (1969) reported that 58 per cent of victims expressed feelings of inferiority. However, Gomes-Schwartz *et al.* (1990) found that pre-school sexually abused children exhibited a more positive self-concept than the normative population. This finding held true for the older age groups as well. Conte and Schuerman (1987) pointed out that victims of CSA develop guilt symptoms.

(ii) *Depression, Withdrawal and Suicide:* Friedrich *et al.* (1986) found from a sample of 61 sexually abused females that 46 per cent were experiencing a range of internalized emotions including depression with two years of being abused. They also found that withdrawn reaction was more common in younger victims. Anderson *et al.* (1995) found that 25 per cent of females who had been sexually abused showed symptoms of depression afterwards. Other studies have also pointed out to an internalizing response by school age sexually abused victims (Glad Stone *et al.*, 1999; Kolko *et al.*, 1988). Lindberg and Distad (1985) evinced that one third of adolescent victims of sexual abuse had attempted suicide.

(iii) *Fearfulness:* Browne and Finkelhor (1986) revealed that the most common initial effect noted in empirical studies is fear. Gomes-Schwartz *et al.* (1990) found 45 per cent of their most vulnerable group of children (7-13 years old) were experiencing fearful reaction to what had happened to them within the first six months following the onset of abuse. However, Beitchman *et al.* (1991) pointed out that similar degrees of fear and depression are to be found in general psychiatric population and therefore the emotion attributed in the studies specifically to sexual abuse could be linked to other stress including factors. Apart from this, child sexual abuse has certain long-term consequences.

Consequences Associated with Psychological Maltreatment

Both researchers and professionals have identified the potential short- and long-term consequences of psychological maltreatment, as depression, withdrawal and low self-esteem (e.g., Gross & Keller, 1992). In the interpersonal realm, researchers have documented maladjustment in psychologically maltreated children in the areas of attachment, social adjustment, and peer relationship. Psychologically maltreated children, for example, are significantly more likely than their non-maltreated peers to be insecurely attached to a parent (Egeland, 1991; Crittenden & Ainsworth, 1989). Several investigators have also found that psychologically maltreated children exhibit lower levels of social competence and adjustment relative to their non-maltreated counterparts (e.g., Brassard, Hart & Hardy, 1991; Claussen & Crittenden, 1991; Vissing *et al.*, 1991). However, a limited number of studies have evaluated the potential long-term effects of psychological maltreatment. The available information indicates that adults who report a childhood history of psychological maltreatment exhibit psychological difficulties. Gross and Keller (1992) evaluated 260 university students, identified as physically and psychologically abused. Respondents had lower self-esteem scores than non-abused respondents but did not differ significantly on measures of depression and attribution style. Respondents experiencing both physical and psychological abuse, however, exhibited higher levels of depression than non-abused college students or college students with a history of only one type of abuse. Regression analysis also revealed that psychological abuse was a more powerful predictor of depression, self-esteem, and attribution style. Other studies have confirmed the presence of low self-esteem, anxiety, depression, dissociation and interpersonal sensitivity in adults with a history of psychological maltreatment as well as the central role of psychological maltreatment in predicting the effects of child maltreatment (e.g., Kent & Waller, 1998; Downs & Miller, 1998; Briere & Runtz, 1988;1990; Hoglund & Nicholas, 1995). Present researcher thus planned to study the outcome of child abuse and neglect on personality development of children.

THE PRESENT STUDY

Child maltreatment (abuse and neglect) is a widely studied area of psychological investigation particularly in developed countries. Findings of these studies have not only explored the deep-rooted causes of child maltreatment but also have significance for preventive interventions. In larger number of researches, the psychological consequences of child abuse have been studied in relation to classificatory or structural factors and therefore, causal explanations become little difficult. Apart from this, certain methodological problems in these studies are also observed. Firstly, most of the studies are based on secondary data (either registered or reported samples) and another noticeable point is, that these studies are mainly conducted in developed countries, where multifaceted preventive efforts at multiple levels have been proliferated since the last 4 to 5 decades. However, this picture is quite different in developing countries like India. Because of the contextual and cultural variations, the nature and forms of abuse will differ in developing and developed countries, thus, studies suffer from correct screening of abused cases and adequate research methodology. Although, child maltreatment is present at alarming volume, but correct statistics of cases are unknown due to negligence, lack of awareness and fear of defame. Some forms of abuse also vary in Indian context as compared to western countries. Although some efforts by sociologists (Chattoraj & Saxena, 1990; Mehta, 1979) and medical professionals (Bhattacharya, 1979) have been made to investigate the forms of maltreatment, however, these studies provide some indication/clue, to understand the socio-demographic and pathological causes of abuse and neglect, yet, in order to understand the psychological causes and consequences of abuse and neglect, the role of process variables should be properly investigated. Hence, it was considered important to conduct a psychological study on child abuse and neglect in relation to numerous socio-demographic and psychological variables. Furthermore, various psychological processes and personality characteristics have been studied in relation to child abuse and neglect, in present research.

The specific research problems, studied in relation to child abuse and neglect are :

(i) Child Abuse and Neglect in Relation to Contextual and Socio-Demographic Factors

Apart from extensive studies conducted on child abuse and neglect throughout the world, there are still numerous questions unanswered concerning the role of contextual and socio-demographic factors in the origin of child abuse and neglect and their interplay in the development of children.

(ii) Parental Factors and Child Maltreatment

Researchers have established an exclusive relationship between certain adult characteristics and child maltreatment. Adults, who were abused as child, were found more abusive than control group (Kempe & Helfer, 1972). Apart from this, maltreating parents exhibited cognitive deficiencies (Wolfe, 1985), high anger and high arousal level (Averill, 1983) and other emotional imbalances (Bower, 1981) than non-abusive counterparts. However, several other evidences do not support this and suggest that there must be certain other factors adding to the origin of abuse and neglect. Hence it was found relevant to investigate the role of parental factors in child abuse and neglect.

Researches revealed that negative mental states were found to be positively linked with child maltreatment. Stress may be the cause of maltreatment or *vice versa*. Certainly, child abuse causes overwhelming amount of stress, especially when the attack was unexpected. Thus, a sizeable number of researches attempted to determine the relationship between stress and health/outcomes.

Apart from extensive studies in the field of child abuse and neglect, researchers observed some of the unexplored dimensions of child abuse and neglect especially in Indian context.

(1) Child abuse and neglect in relation to contextual, social and demographic variables, such as:

(a) Rural-Urban environmental setting,
(b) Social status in terms of caste and occupation, and
(c) Demographic factors in the origin of child abuse and neglect.

(2) Forms of Child abuse, such as:

(a) Child abuse within the family,
(b) Children as domestic servants,
(c) Children employed by owners of tea stall, *dhabas* and hotels,
(d) Child abuse by adults who are socially derailed, loose in characters or mentally sick, and
(e) Child maltreatment by sick parents.

(3) Psychological consequences of child abuse and neglect.

Objective

On the basis of queries raised above, this research was carried out with the following objectives:

1. To find out the demographic variables which provide germane grounds for child abuse and neglect in rural-urban societies of eastern districts of Utter Pradesh,
2. To examine the socio-economic status and occupational characteristics of families in which children are abused and neglected,
3. To investigate the psychological dispositions of abusive parents, and
4. To measure the Cognitive, Motivational and Behavioral functioning, and personality pattern of abused and non-abused children.

Keeping these objectives in view, present study was carried out to answer following research question :

1. Whar are the extent and forms of child abuse and neglect in rural-urban socities?
2. Which of the demographic factors provide germane grounds to the incidence of child abuse and neglect? In other words, whether there is any difference at the level of abuse and neglect with varying demographic characteristics of parents, children and family?
3. What are the patterns of relationship between demographic characteristics (of children, parents and families) and child abuse/neglect?
4. Which of the demographic factors contribute to the incidence of child abuse and neglect in rural-urban societies?
5. What are the extent and forms of child abuse and neglect across different social status and occupational characteristics of the families?
6. In what ways, are social status and occupational characteristics of families related with child abuse and neglect?
7. How do social status and occupational characteristics of parents predict child abuse and neglect?
8. Whether there is any difference on psychological dispositions of abusive and non-abusive parents? In other owrds, whether the pattern and level of parental attitude, strees, anxiety and coping responses are different in abusive and non-abusive parents?
9. How are psychological characteristics of parents related with child abuse and neglect?
10. How do psychological characteristics of parents contribute to child abuse and neglect?
11. What are the consequences of child abuse and neglect on cognitive, motivational and behavioural functioning, and personality pattern of children? More specifically;

(a) Whether abused children perform inferior on various cognitive functioning, viz. perceptual identification, verbal organization, memory span and intellectual tasks than non-abused counterparts. Whether, these functioning also vary across different residential settings, gender, age, and socio-economic status of children?

(b) Whether levels of achievement and aspiration differ in accordance with varying levels of child abuse, residential setting, gender, age and SES of children?

(c) Whether levels of aggression and helping behaviour are different in accordance with varying level of child abuse, residential setting, gender, age and SES of children?

(d) Whether the personality patterns of children also vary across different levels of abuse and age of children?

12. In what ways, are child abuse and neglect related with different levels of cognitive, motivational and behavioural processes and personality patterns of children?
13. How do child abuse and neglect contribute to cognitive, motivational and behavioural functioning and personality pattern of children?

2

METHOD

Plan and Strategy of the Study

In this chapter, despite the methodological problems involved in the study of child abuse and neglect, the plan and strategy of present research have been discussed. For the screening of abused and neglected cases and to finalize the measures of abuse and neglect, a preliminary study was conducted. Apart from this, an attempt has also been made to present an overview of universe and sample characteristics, research strategies of antecedent, mediating and outcome variables, selected in the proposed study.

Research Strategy

The dynamics of child abuse and neglect are multifaceted concerning environmental, demographic, socio-cultural and psychological factors. Child abuse and neglect may take place at multiple points as a result of multi-causative factors. Belsky (1980) has conceptualized, "Child abuse and neglect" as a Social-Psychological Phenomenon and indicated that multiple forces determine it across four different levels; in the individual (ontogenetic development), the family (the micro-system), the community (the exo-system), and the culture (the macro-system), in which both the individual and family are embedded.

The agents of child abuse and neglect (Perpetrators) are found within and outside the family. Significant perpetrators may be parents, family members, relatives, neighbours, peer group, school-teachers and any significant member for the child. The incidence rates of child abuse and neglect in eastern districts of U.P. are observed in plenty; however, correct statistics of the abused and neglected cases are either over or under estimated due to methodological problems in the study of child maltreatment. Therefore, multiple criteria were adapted for the screening of cases: (i) Self report measures: Two measures, i.e., child abuse checklist and child neglect scale developed by Pandey, S.

(2002) were used to identify the abused and neglected cases. (ii) Reported/registered cases: Registered cases of child abuse and neglect or cases reported by adults, viz. school teachers, pediatrics, health professionals, or neighbours were considered as secondary data. Afterwards these self-report measures were administered to determine the intensity and frequency of abuse and neglect.

Subject selection procedure was used for categorizing the level of abuse neglect & settings and other characteristics like gender, age and socio-economic status of respondents. According to D'Amato (1970), "In an experimental study, use of classificatory factor is an accepted technique of experimentation". Calfee (1975) also confirms that the major experimental variable has to be considered as classificatory factor. This procedure is extensively exercised in a sizeable number of psychological researches (Misra & Tripathi, 1980; Berry, 1976). Since the participants of present study were children (6-18 years) and parents (one of the parents, whoever was reported more abusive), thus some of the variables were investigated on children, while certain other variables were investigated on parents. Variables and concerned measures, administered on children are displayed in Table 2.1.

Table 2.1: Shows variables along with their measures administered on children

S. No.	*Variables*		*Measures*
1.	Child abuse		Child Abuse Checklist (S. Pandey, 2002)
	Child neglect		Child Neglect Scale (S. Pandey, 2002)
2.	**Environmental Factors:**		Personal Data Sheet (PDS) (S. Pandey, 2002)
	Rural-urban setting		Personal Data Sheet (PDS)
	Family Structure		*Family Environment Scale*
	Family Environment		(M.C. Joshi & R. Vyas, 1997)
3.	Social-status; (caste, socio-economic status), Child-age, gender, education, birth order, health status etc.		Personal Data Sheet (PDS) (S. Pandey, 2002)
4.	**Consequence Variables:**		
	(1)*Cognitive Processes: Intelligence	*	Culture Fair Intelligence Test, (R.B. Cattell: Scales; 1st—1950, 2nd & 3rd—1973).
	Perceptual Identification	*	Embedded Figure Test (E.F.T.) (D. Sinha, 1984)

(Contd...)

(Contd...)

Verbal Learning (clustering)	* Verbal Learning Task (S. Pandey, 2002)
Memory	* Memory Span Test (S. Pandey, 2002)
Creative Thinking	* Verbal and Non-verbal Tests of Creative Thinking. (B. Mehdi, 1973)
(2)***Motivational process:**	
Achievement level	* Achievement Motivation Test (Prayag Mehta, 1978)
Aspiration level	* Aspiration Scale (S. Pandey, 2002)
(3)* Personality pattern:	* Children Personality Questionnaire (C.P.Q.), (R.B. Cattell, 1972). * High School Personality Questionnaire (H.S.P.Q.), R.B. Cattell, 1973.
5. **Behavioral process:**	
Aggression	Aggression Questionnaire (G.C. Pati, 1976)
Helping behaviour	Helping Behavior Questionnaire (S. Pandey, 2002)

Similarly, factors studied on Parents are presented in Table 2.2, systematically.

Table 2.2 : Displays variables along with measuring tools, administered on parents

Variables	*Measures*
Background Information	* Personal Data Sheet (S. Pandey, 2002)
* Socio-Economic Status, Caste Parent's-Age, Gender, Education, Occupation, Income, Health Status, Marital Status etc.	Personal Data Sheet (S. Pandey 2002).
(i) Child abuse and neglect by Parents/adults	Child Abuse and Neglect: Interview Schedule for Parents (S. Pandey, 2002)
(ii) Attitude towards parents	Mohsin Parent Child Inventory (S.M. Mohsin, 1979)
* Stress	* Life Stress Scale (Modified by S. Pandey, 2002 and based on the Social Readjustment Rating Scale by T.H. Holmes & R.H. Rahe, 1967)
* Anxiety	* State-Trait Anxiety Inventory (H.D. Spielberger and Hindi adaptation by S. Sharma & M. Singh, 1973)
* Coping	* Cope Scale (Carver, Scheiver & Weintraub, 1989)

The Screening of Setting and Selection of Participants.

The setting of the study from where sample was drawn consisted of three districts of Gorakhpur division, i.e. Gorakhpur, Maharajganj and Deoria of eastern Uttar Pradesh. Of course, Uttar Pradesh is ranked one of the most backward states of the country, however; it stands first in population density (Total=16,61,74,612, Male : 8,75,60,236, Female=7,86,14,376) as compared to total population of India (Total= 1,02,70,15,274, Male : 53,12,77,078, Female : 49573816) (Table 2.3a).

The table (2.3a) shows population structure Gorakhpur division i.e.; Gorakhpur, Deonia and Maharaj ganj and other characteristic features of localities from which data were collected (based on 2001 census report). According to 2001 census report, in Uttar Pradesh, birth rate is 23.8 and death rate is 10.3 per thousand. The development level is also very low as compared to other states and per capita income is almost unchanged. The literacy rate is also low in U.P. (Total literacy = 57.36%, Male literacy = 70.23%, Female literacy = 42.98%) as compared to the literacy rate of India (Total literacy =68.38%, Male literacy = 75.85% and Female literacy = 54.16%). The average landholding is very low as compared to other states. The industrial development and productivity are also unsatisfactory. Apart from this, medical and transport facilities are also inadequate. The eastern region of Uttar Pradesh consists of the districts of Varanasi and Gorakhpur divisions. These districts are densely populated. In this area majority of population (92.6%) is rural. Due to the lack of income resources and high poverty level, migration to other states is also very high in these regions. Maharajganj and Deoria districts are less developed but have higher poverty line and backwardness than Gorakhpur. This region was therefore viewed the most suitable setting for the study, where incidences of child abuses is seen in plenty. Present-study was thus, conducted in three districts of Gorakhpur division, viz. Maharajganj Gorakhpur and Deoria. The population structure regarding population density, literacy level, health and development, working forces etc. are presented in Table 2.3a. The physical features and socio-demographic characteristics of the region, suggested the appropriateness of setting for present study. The present study was conducted in three districts of Gorakhpur division, namely Gorakhpur, Maharajganj, and Deoria. The population structure (socio-demographic characteristics) i.e. population density, literacy, workforce, child labour etc. based on 2001 census report, are displayed

Table 2.3(a) : Population structure and other characteristic features of Gorakhpur Division (Gorakhpur, Deoria & Maharajganj Region). *Based on 2001 Census Report.*

Name of Location	*Population*	*Population Density (/Sq. Km)*	*M/F Ratio (F/1000 Males)*	*Literacy Rate*	*Social Composition Caste/*	*Main occupation of Dwellers*	*Distance of School*	*School & its type*	*Development level of the Setting*	*Type of child abuse in these localities*
1	2	3	4	5	6	7	8	9	10	11
1. Gorakhpur (a) Ubran	37,84,720 Male-19,31, 762 Female-18,52,958	1140	959	60.96% Male-76.70% Female-	GEN, OBC, SC of Hindu, Muslim, Sikh, Christian	Labour, Service & Business	–	Primary Public & Govt. Inter-college	Under-Developed	*CPA, Child Labour, *CSA, Exploitation, *CPsyA, Substance Abuse, *CPN, *CEdu N, *CEN
(b) Rural 1. Mahnapur	-	-	-	-	SC, OBC & Muslim	Labour, Farming, Small business & Service	3Km	Nearby Village-Primary, Public & Govt. School	Under-Developed	Child Labour, *CPA, *CPsyA, *CSA, Sub-stance Abuse, *CPN, *C Edu N, *CEN
2. Laxmipur	-	-	-	-	Mostly SC & OBC	Labour, Farming &	2Km	Nearby village-	Under-Developed	Child Labour, *CPA, *CPsyA, *CSA, Sub-stance Abuse, *CPN, * C Edu N, *CEN
3. Bhagatpurava	-	-	-	-	Mostly SC & OBC	Labour, Farming & Small business	2Km	Nearby village-Primary & Middle school	Under-Developed	Child Labour, *CPA, *CPsyA, *CPN, *C Edu N, *CEN
4. Stenpur Bhatolia	-	-	-	-	SC	Mostly Labour & Small business	2Km	Nearby village-Primary & Middle school	Under-Developed	Child Labour, *CPA, *CPsyA, Substance Abuse, *CPN, *C Edu N, *CEN

Contd...

Table 2.3(a) : Contd...

II. Deoria (a) Urban	27,30,376 Male-13, 63,250 Famele-13, 67, 126	1077	1003	59.84% Male-76.31% Female-43.56%	GEN, OBC, SC of Hindu and Muslim	Labour, Service Farming & Business		Primary, Convent & Govt. school & college	Under-Developed	Child Labour, *CPA, *CSA, Substance Abuse, *CPN, *C Edu N, *CEN
(b) Rural 1. Badhauna	-	-	-	-	Mostly GEN, OBC & SC	Labour, Service & Farming	3Km	In village itself Primary school near by village -Primary & inter-college	Under-Developed	Child Labour, *CPA, *CPsyA, *CSA, *CPN *C Edu N, *CEN
(2) Andila	-	-	-	-	Mostly GEN, OBC & SC	Labour, Service & Farming	2Km	Primary School & Nearby village inter-college	Under-Developed	Child Labour, *CPA, *CPsyA, *CSA, Substance Abuse, CPN, *C Edu N, *CEN
(3) Deobari	-	-	-	-	Mostly GEN, OBC & SC	Labour, Service & Farming	1.5 Km	Nearby village Primary & Govt. Inter-college	Under-Developed	Child Labour, *CPA, *CPsyA, *CSA, Substance Abuse, *CPN,
(4) Kothi	-	-	-	-	Mostly GEN, OBC, & SC	Labour, Service & Farming	5 Km	Primary school in village & nearby village Primary school & inter-college	Under-Developed	Child Labour, *CPA, *CPsyA, *CSA, Substance Abuse, *CPN,
(5) Mauna	-	-	-	-	Mostly GEN, OBC & SC	Labour, Service & Farming	0.5Km	In village Primary & Govt. inter-college	Under-Developed	Child Labour, Child, Trafficking, harassment, *CPA, CPsyA, Substance Abuse, *CPN, *C Edu N, *CEN

Contd...

Table 2.3(a) : Contd...

III. Deoria (a) Urban	21,67,041 Male-11,20,800 Female-10,46, 241	734	933	47.72% Male- 65.40% Female- 28.64%	GEN, OBC, SC of Hindu & Muslim	Labour, Service, Business & Farming	-	Primary, Public & Govt. school & college	Under-Developed	*CPA, *CPsyA, *CSA, *CPN, *C Edu N, *CEN
(b) Rural (1) Paragpur	-	-	-	-	GEN, OBC, SC of Muslim	Labour, Farming & Business	1 Km	Nearby village Primary, Convent & Govt. inter-college	Under-Developed	Child Labour, *CPA, *CPsyA, *CSA, Sub-Abuse, *CPN, C Edu N, *CEN
(2) Harkhapas	-	-	-	-	GEN, OBC & SC	Labour, Farming & Business	3 Km	Nearby village Primary, Convent & Govt. inter-college	Under-Developed	Child Labour, *CPA, *CPsyA, *CSA, Substance Abuse, Harassment, *CPN, *C Edu N, *CEN

**CPA*= Child Physical Abuse; **CSA*=Child Sexual Abuse; **CPsyA*= Psychological Abuse;
* *CPN*= Child Physical Neglect; **C Edu N*= Child Educational Neglect; **CEN*=Child Emotional Neglect.

in Table 2.3a. This table indicates, sharp variation on various characteristics among three districts. Gorakhpur stands first on population density (Total population = 37,84,720, Male population= 19,31,762 and Female population = 18,52,958). Deoria comes on second order (Total population = 27,30,376, Male population= 13,63,250 and Female population = 13,67,126) and Maharajganj is on third order (Total population = 21,67,041, Male population= 11,20,800 and Female population= 10,46,241). Districts differ on literacy rate, Gorakhpur (44.48%) and Deoria (43.56%) have higher literacy rate than that of Maharajganj (28.64%) which is much low as against total literacy percentage of Uttar Pradesh (57.36%). The majority of population in these districts is rural and illiterate. The other characteristics of rural-urban settings are not homogenous. The rural population is deprived of various facilities, viz., schooling, communication channels, transportation facilities, health centres and sources of work forces whereas, urban setting enjoys all these facilities. However, no sharp variation was found regarding composition of caste, community, and child labor. Therefore, these features provided appropriate setting for the study of child abuse and neglect. The geographical setting of these districts was also appropriate for present investigation as cases of gender discrimination, girl child harassment; child trafficking and child labor have been observed in plenty.

Participants

A total of 335 families (child and one of the parents whoever were reported more abusive) were randomly selected from rural and urban areas of three districts of eastern U.P. Stratified random sampling technique was exercised for sample selection. The number of cases drawn from different groups is displayed in Table 2.3(b).

Two strategies were adopted to screen the cases of abuse and neglect. (i) Reporting by neighbors, other family members, health professionals, schoolteachers and significant others in the form of secondary data. Then, (ii) Child abuse checklist and child neglect scale were administered on children to determine the level of abuse and neglect. Median points for child abuse (Mdn = 123) and neglect (Mdn = 25) were determined on the basis of responses given by children on both measures.

On the basis of median scores, children were divided into abused/ non-abused groups. It is evident from the Table 2.3 (b), a total of 175

Bhojapuri to solve the problems of communicating instruction and items in understandable terms. In present study various measures were selected or developed to fulfil the purpose of this research investigation. Certain measures were applied for parents whereas a set of measures was administered on children. A brief description of these measures is given below.

Measures administered on parents

1. *Personal Data Sheet (P.D.S.)*

In order to determine the socio-demographic and familial characteristics of the parents/adults Personal Data Sheet (Pandey, S., 2002) was used.

2. *Child Abuse-Neglect Interview Schedule for Parents (CANISP)*

To assess the perception and practices of child abuse and neglect by parents, this interview schedule (Pandey, S., 2002) was used. The schedule includes two sections, i.e. A and B. The items of section A are open ended questions and respondents were free to answer these questions in their own statements, hence, such responses required qualitative analysis. Contrary to this, section 'B' contained closed end items and respondents had to put (√) mark on any one of the alternatives. Marked responses were counted for further analysis. This interview schedule is pilot tested and has also been used in many studies (Pandey, 2002, 2004).

3. *Mohsin-Parent-Child Inventory*

In order to determine the attitude of parents towards their own parents, Mohsin-Parent-Child Inventory (1979) was used. This inventory is generally called as parental attitude inventory, is intended to be used as indirect measure of the respondent's attitude towards his/her parents. Thus, to know whether parents had negative or positive relationship and attitudes about their parents, this inventory was used. This inventory consists of 50 statements, 25 conveying permissive and 25 restrictive disciplinary practices. It is a, 4-point scale ranging from very true to very untrue. High scores signify positive (permissive) and low scores denote negative (restrictive) attitudes. Thus, it is inferred that low scorers, having restrictive attitude towards their parents, might be the parents who were abused/neglected as children. This is a highly reliable (split half, r =.76), (test-retest, r =.70) and valid (construct validity, r =.39) measure.

Scoring : The items of the Mohsin-Parent-Child Inventory are to be checked on a 4 = point scale: very true, true, untrue, very untrue, in the order against a restrictive (negative) statement; and in the reverse order against a permissive (positive) statement. In a positive item very true is scored as a 4, True as 3, untrue as 2 and very untrue as 1, and in a negative item, very untrue is scored 1, untrue as 2, true as 3 and very true as 4. The maximum score possible is 200 and minimum is 50.

4. *Life Stress Measure*

This is a modified version of Holmes and Rahe Readjustment Scale (1967). This measure consisted of 21 items pertaining to stressful events. This scale was used to measure the feeling of stress in parents. Respondents were asked to respond on a 5-point scale ranging from very much (1) to not at all: (5). The scoring was done following 5, 4, 3, 2 and 1 order and total summated scores indicated magnitude of stress in parents. Reliability of this measure was found high (r =.76). High scores indicated high level of stress in respondents whereas low scores denoted low level of stress in parents.

5. *State-Trait Anxiety Inventory*

To assess the level of anxiety in parents, State-Trait Anxiety Inventory, developed by Spielberger (1970) and translated by Sagar Sharma, (1973) was used. High scores denoted high level of anxiety, whereas, low scores referred to low level of anxiety in parents.

6. *Cope Scale*

To determine the coping styles and strategies used by parents to confront stressful life events, Cope scale (Carver, Scheiver & Weintraub, 1989), was used. This scale includes 15 domains of coping strategies related to active, adaptive and maladaptive styles. The scale consists of 60 items and respondents had to answer each item on 4 point scale. The reliabilities for the fifteen sub dimensions are: Active coping r =.68, Planning r =.73, suppression of competing activities r =.63, positive reframing r =.64, restraint coping r =.62, use of instrumental support r = .64, use of emotional support r = .71, religion r =.82, acceptance r = .57, focus on and venting of emotion r =.50, denial r =.54, mental disengagement r=.71, behavioural disengagement r = .65, humor r =.64 and drugs r =.65.

Scoring: The scoring was done for each subscale separately. Responses given on item number related to each dimensions, were added and state of coping responses on each domain was determined.

Measuring Tools, Administered on Children

1. *Personal Data Sheet (P.D.S.)*

In order to determine the socio-demographic and familial characteristics of children, personal data sheet was used.

2. *Child Abuse Csshecklist*

This checklist was developed and standardized by Pandey, S. (2002) to determine the child's physical, psychological and sexual abuses and abuse as a whole, separately. Initially this checklist contained 100 items related to physical, sexual and psychological abuses by parents/family members or any significant others and children had to respond on a three point scale, given in front of each items.

In pilot study, try out was done to gain insight about the appropriateness of language and suitability of the items. A close perusal of the responses given on items suggested that some of the items were either irrelevant or ambiguous, therefore, these items were dropped out from the final set and ultimately 75 items were retained in the child abuse checklist (shown in Table 3.5). The reliability (test-retest r = .69, reliability index r = .89) and validity (criterion validity r= .83) of the checklist were found high. This checklist is extensively used in other studies (Pandey, S., 2002, 2004).

Scoring: Responses on each item were scored following 3, 2 and 1 order. Total summated scores in each area denoted the magnitude of abuse in specific area and on the basis of the grand total, abuse, as a whole was determined.

3. *Child Neglect Scale*

In order to determine the neglectful experiences received by parents, concerning physical, emotional and educational areas as well as neglect as a whole. Child neglect scale, developed and Standardized by Pandey, S., (2001) was used. This scale initially contained 25 items related to physical, emotional and educational neglects. This is a three point's scale. Try out of the scale was done to determine the appropriateness of the language, format and suitability of the items. It was found that some of the items in the scale were either irrelevant or ambiguous therefore, those items were rejected and finally a total of 15 items were finalized in child neglect scale for the main study. The reliability (r=.91) and validity (criterion r=.79) of the scale were found high.

4. *Family Environment Scale*

To identify the family environment of the different groups of children, Family Environment Scale was used. This scale is constructed in Hindi language by Joshi and Vyas (1997) to assess the perception of family environment by children. Family environment scale contains a total of 79 items concerning ten sub scales, i.e. cohesion, expressiveness, conflict, independence, achievement orientation, intellectual-cultural orientation, active recreational orientation, moral-religious emphasis, organization and control. The reliability (r value for subscales ranged from .58 to .88) and validity (r value for subscale ranged from .28 to .41) of the scale are found significant. Children were asked to respond on 5-point scale ranging from 0 to 4.

Scoring : Positive items were scored following 4, 3, 2, 1 and 0 order whereas, reverse pattern was followed in scoring of the negative items. The total summated scores for each domain represented the magnitude of family environment for children.

Measures of Outcome Variables

Various measuring tools and tasks were used to study the impact of child abuse on cognitive, motivational and behavioural functioning and personality pattern of children. A brief description of these measures is given below :

(i) Cognitive Measures

(a) *Embedded Figure Test:* To ascertain the perceptual identification by different groups of children, Sinha Pictorial Embedded Figure Test was used. This test is a highly reliable (r = .61) and valid (r = .70) measure.

Scoring: Scoring of the each stimulus, which is successfully located, is given a score of one and the total score on the test is the total number of stimuli located by the subject, maximum possible scores are 41 and minimum is zero.

(b) *Verbal Learning Task:* In order to study the pattern of organization (clustering) in verbal learning, two sets of verbal task were devised. First list included 15 meaningful words related to four categories, whereas items of second list were unrelated to each other and were of different concepts. Each child was presented the list of items one by one and was asked to respond in any order. Thus, free recall method was used for the study and clustering pattern was determined on the basis of obtained responses

Scoring: The success of each item was given a score of one (1) and failure, a score of zero (0). Thus, maximum score on the test is 15 and minimum is zero.

(c) Memory Span Test: To ascertain the immediate memory span of children, a set of 7 numbers, varying 3 to 7 digits were devised. After the presentation of each digit, children were asked to recall the number. Correctly recalled items denoted the memory span of the children.

(d) Tests of Creative Thinking: To determine the creative potentials in children, the verbal and non-verbal tests of creative thinking (developed and standardized by B. Mehdi, 1973) were used. The level of verbal fluency, flexibility and originality were assessed by verbal test of creative thinking and non-verbal test was used to determine, elaboration (Pictorial and Title), originality (Pictorial and Title) and flexibility in children. Both tests are highly reliable and valid. The reliability (r = .946) and validity (r = .385) of non-verbal test are reported high. Similarly, verbal test of creative thinking is also found highly reliable (total creativity .959) and valid (total creativity = .959).

(e) Culture Fair Intelligence Test (C.F.I.T.): In order to assess the intellectual capacity of children, C.F.I.T. was used. R.B. Cattell developed this test to measure general intelligence across the culture. It consists of the task requiring basic cognitive skills involving complex conceptualization. Since the age of children ranged from 6-18 yrs, hence 1st, 2nd and 3rd scales of this test were used. The administration of the test was done in accordance with the procedure given in the manual. The participant 's' score was the number of correct solution done by him/her.

(ii) Motivational Measures

(a) Achievement Motivation Test: To study of achievement level in children, the Achievement Motivation Test (developed by Prayag Mehta, 1978) was used. This test contains six T.A.T. type pictures and children had to write stories on each card. Each story was content analyzed, according to the rules given in the manual. This test is a highly reliable (test-retest: r = .72, split half: r = .54) and valid (r = .44) measure of achievement.

Scoring : Stories are scored in a standard way. First, each story is scored for the presence of achievement imagery following the criteria given in the manual. Therefore, stories scored for achievement imagery are taken up for further scoring for each of sub-categories.

(b) Aspiration Scale : To assess aspiration level of children related to seven different areas (viz., economic, security, power, prestige/self-esteem, altruism and self-actualization), a ten point-ladder scale previously used by Trommsdroff, 1988) was used. Children were asked to put (Ö) mark to indicate their status of aspiration, on 10-point ladder scale. The first step or rung of the ladder indicated lowest status and 10th step or rung denoted highest status of aspirations. Such ladders have been successfully used by a sizeable number of researchers to study hopes and fears of respondents (Trommsdroff, 1988; Mishra & Tripathi, 1980; Muthaya, 1971; Cantrail 1965).

(iii) Behavioral Measures

(a) Helping Behavior Questionnaire: In order to assess helping behaviour in children, Helping Behavior questionnaire (Pandey, S., 2002) was used. The questionnaire contains eight items describing various helping situations and subjects had to respond on 5 point scale by putting (Ö) mark given in front of each items. Test-retest reliability of the questionnaire is found high (r = .88). The criterion validity of the questionnaire is also found highly significant(r = .87). High scores denoted more helping behavior whereas; low scores, indicated low helping tendency in children.

(b) Aggression Questionnaire: To determine the level of aggression in children, aggression questionnaire developed by G.C. Pati (1976) was used. This scale includes 16 situations with 3 points closed-ended scale. The responses obtained on each item are transformed into standard score. Higher scores denoted more aggression. The reliability and validity of the test are reported high.

(iv) Personality Questionnaire

To check the personality pattern of abused/neglected and normal children, Cattell's C.P.Q. (Children Personality Questionnaire) and H.S.P.Q. (High School Personality Questionnaire) were used to measure personality pattern of the different age group of children. The C.P.Q. was applied on 8-12 yrs. age group of children. The reliability (ranged from r = .42 to .75) and validity (r value ranged from .75 to .90) were found to be high. Further, H.S.P.Q. was administered on 13-18 yrs. of children. The reliability (r value ranged from .51 to .80) and validity of this test (r values range from r = .57 to. .74 respectively) are high. C.P.Q. contains 140 items, 70 in each forms (B1 & B2) related to fourteen factors of personality. Children were asked to respond on one of the two alternatives.

Similarly H.S.P.Q. includes 142 items and children were requested to put (Ö) mark on any one point of the three alternatives, i.e. a-b-c. Obtained scores, known as raw scores, were then transformed into Sten scores for each factor.

Procedure

Since the screening of geographical settings of the study and identification of cases were already done in preliminary phase of the study, thus on the basis of secondary data, the research team consisting of principal investigator and project fellow approached to the respective areas of the study and tried to establish rapport with the family members. This study was conducted in rural and urban area of Gorakhpur division, viz. Gorakhpur, Maharajganj and Deoria districts; thus, research team visited these places one by one. Prior to this, a brief orientation session, to introduce the nature and purpose of the study and detailed instruction about administration of measuring tools, was organized by principal investigator. High degree of skill and ingenuity on part of the investigator was essential to communicate the respondents in effective manner. The sample of this investigation consisted of children and their parents, thus, multiple phases for data collection were required. Since, availability of subjects for the study was a difficult problem because sample characteristics differed on degree of environmental setting (rural-urban), age (6-10, 11-14 and 15-18 yrs.), gender (boys-girls), socio-economic status (high, middle and low) and level of abuses. So, it was quite difficult to approach the respondents and schedule them for administration of the measuring tools yet, efforts were made to contact the children and their parents in different settings. The study was conducted into two phases.

First Phase of the Study

First of all, children were contacted for this investigation. This phase required multiple sessions.

1st Session: In this session, children were contacted in the school or at work place (in case of child labor). They were explained about the nature and purpose of the study in effective manner and proper rapport was established, then, they were given a booklet containing Personal Data Sheet (P.D.S.), Child Abuse Checklist and Child Neglect Scale. Children were requested to respond on these measures one by one. As they completed responses, data were collected and they were thanked for participation and further, were requested to participate in 2nd session of the study. First session ended with administration of child abuse and neglect scales. Other tests and task were administered in 2nd and 3rd sessions.

2nd Session: In this session, subjects were given experimental tasks, tests and questionnaires. First of all children were given different types of cognitive measures, i.e. Embedded Figure Test, Task of Verbal Organization (clustering) and Memory Span. They were requested to perform on these tasks one by one. As, they responded on these tasks; they were given verbal and non-verbal tests of Creative Thinking and Culture Fair Intelligence Test respectively. After completing the responses on these measures, children were thanked for co-operation and again requested to participate in 3rd session of the study.

3rd Session: In this session, first of all achievement and aspiration levels of children were planned to assess. Hence, they were given Achievement Motivation Test and Aspiration Scale and requested to respond on these measures. Then, to ascertain the aggression levels and helping behaviour, children were requested to respond on Aggression Questionnaire and Helping behaviour questionnaire subsequently. Lastly, CPQ/HSPQ was administered to determine the Pwesonality Pattern of children. As they completed responses on various said measures, data were collected and subjects were thanked for co-operation.

Second Phase of the Study

In this phase of the study, parents were contacted either at home setting or work place and requested for their co-operation. After getting their consent, they were described about the nature and aim of the study. Then, they were given child abuse and neglect interview schedule, Mohasin-Parent-Child-Inventory, Life-Stress Scale, Strait-Trait Anxiety Scale and Cope Scale subsequently.

They were requested to respond on each measure one by one. As they completed their responses on above-mentioned measures, data were collected and parents were thanked for participation.

Data obtained on various measures were then scored according to rules given in manuals. Scores were treated statistically in terms of Comparātive analysis (Analysis of Variance & t test), Correlational analysis and Stepwise Multiple Regression Analysis (SMRA). Present results are displayed and interpreted in accordance with objectives of the study, in result section.

3

RESULTS

I

Child Abuse and Neglect in Relation to Ecological and Demographic Characteristics

First objective of the study was to find out the demographic variables, which provide germane grounds for child abuse and neglect in Rural and Urban societies of eastern districts of U.P. The main findings of the present study indicated that the magnitude and patterns of child abuse and neglect varied significantly in rural-urban societies. Apart from this, certain demographic and social characteristics, contributed significantly in the origin of child abuse and neglect. Obtained results are interpreted in this section:

(i) Child Abuse and Neglect in Rural and Urban Societies

In order to determine the patterns and salience of abuse and neglect in rural and urban societies, ANOVA analysis was done. Results displayed in Table 3.1 revealed that various forms of abuse and neglect varied significantly in rural-urban societies. Rural children were found more physically abused, than urban children. They were also found more emotionally neglected than urban children. Educational neglect was also found higher in rural children than urban children. Neglect as a whole, was also found greater in rural children as compared to urban counterparts.

(ii) Child Abuse and Neglect in Relation to Demographic Characteristics

Present results evinced the role of demographic characteristics of child, family and parents in the origin of child abuse and neglect. Results are displayed in tables and interpreted in nutshell :

Table 3.1 : Shows means, SDs and F-values of child abuse ánd neglect across rural-urban setting

Residential Setting	*Abuse*				*Neglect*			
	Physical Abuse	*Psychological Abuse*	*Sexual Abuse*	*Abuse as a whole*	*Physical Neglect*	*Emotional Neglect*	*Educational Neglect*	*Neglect as a whole*
Rural								
Mean	42.68	74.16	9.36	126.52	4.01	17.35	5:01	26.37
S.D.	7.34	13.01	2.22	19.37	1.38	3.69	1.84	6.00
Urban								
Mean	40.86	72.47	9.44	122.74	3.86	16.42	4.55	24.85
S.D.	6.62	13.37	2.63	19.35	1.42	3.47	1.54	5.72
F- values	5.73*	1.37	0.08	3.17	0.87	5.61*	6.39**	5.67*

** P < .01

* P < .05

(i) Incidence of Child Abuse and Neglect by the Child Characteristics

Certain child characteristics like age, gender, education and health status were found as potential factors, contributing to child abuse and neglect.

1. *Child's Age:* Results indicated a consistent picture of age differences towards the incidences of abuse and neglect. Child abuse varied significantly across different age groups. Preadolescents were found more physically abused (M= 43.28) than younger children (M= 42.39) and adolescents (M= 39.17), F (2,332) = 11.11, P<.01. Contrary to this, sexual abuse was found higher in adolescents (M= 10.21) than pre-adolescents (M= 9.44) and children (M= 8.58) respectively, F (2,332) = 12.76, P<.01.
2. *Child's Gender:* The ANOVA results revealed that boys and girls differed significantly on Physical neglect [F (1,333) = 7.92, P<.01], Educational neglect [F (1,333) = 8.77, P<.01] and Neglect as a whole [F (1,333) = 5.90, P <.05]. More specifically, girls were found more neglected [M (phy. neg.] =4.15, M, (edu. neg.) =5.04, and M (neg. as a whole) =26.34] than the boys [M (phy. neg.) =3.72, M (edu. neg.) =4.99, and M (neg. as a whole) =24.78].
3. *Child's Education:* Results evinced that various forms of abuse and neglect differed significantly across different education level of children. As Table 3.2 indicates that illiterate/low educated (0-5th class) children reported high levels of Physical abuse, Psychological abuse, Abuse as a whole, Physical neglect, Emotional neglect, Educational neglect and Neglect as a whole than highly educated children.
4. *Child's Birth-Order:* It is apparent from the result that the educational neglect differed significantly across birth order of the children F (3,331) = 3.01, P<.01. Later born children were found more educationally neglected (M= 6.11) than earlier born (M= 4.24).
5. *Child's Health Status :* Table 3.3.denotes that health status of the children is also found an important contributing factor of Child Abuse and Neglect. Children having poor health status reported maximum physical abuse, psychological abuse, abuse as a whole, emotional neglect, physical neglect,

Table 3.2 : Shows Means, SDs and F-values of child abuse and neglect across different education level of children

Children's Education	*Abuse*				*Neglect*			
	Physical Abuse	*Psychological Abuse*	*Sexual Abuse*	*Abuse as a whole*	*Physical Neglect*	*Emotional Neglect*	*Educational Neglect*	*Neglect as a whole*
0-05th Class								
Mean	43.87	75.50	9.20	128.44	4.27	17.28	5.13	26.69
S.D.	7.65	13.62	2.44	20.82	1.54	3.56	1.82	6.09
06-08th Class								
Mean	40.40	72.36	9.43	122.84	3.64	16.81	4.55	25.01
S.D.	5.21	12.29	2.36	16.83	1.22	3.43	1.68	5.53
09-10th Class								
Mean	38.52	69.61	9.66	117.95	3.55	16.06	4.18	23.79
S.D.	5.45	12.87	2.41	17.19	1.08	3.79	1.21	5.43
11-12th Class								
Mean	36.80	67.10	10.80	114.70	3.10	15.10	4.10	22.30
S.D.	4.13	3.36	3.26	11.54	0.32	2.89	0.88	3.68
F- values	14.58**	4.45**	1.75	6.40**	7.84**	2.81*	6.67**	5.77**

*P < .05
** P < .01

Table 3.3 : Shows Means, SDs and F-values of child abuse and neglect in accordance with health status of children

Children's Health	*Abuse*				*Neglect*			
Status	*Physical Abuse*	*Psychological Abuse*	*Sexual Abuse*	*Abuse as a Whole*	*Physical Neglect*	*Emotional Neglect*	*Educational Neglect*	*Neglect as a Whole*
Handicapped								
Mean	36.75	74.50	8.50	119.75	3.00	17.25	4.75	25.00
S.D.	3.59	1.92	1.00	4.92	0.00	2.06	1.50	3.46
Weak								
Mean	49.80	78.10	9.25	137.15	4.70	18.70	5.35	28.75
S.D.	8.79	12.47	2.63	20.52	1.49	2.85	1.69	5.22
Normal								
Mean	42.02	74.83	9.56	126.62	4.04	17.22	4.98	26.25
S.D.	6.82	12.84	2.63	18.93	1.46	3.52	1.74	5.85
Excellent								
Mean	38.29	65.79	8.94	112.86	3.35	14.85	3.77	21.97
S.D.	4.51	12.62	1.57	16.24	0.89	3.41	1.11	4.79
F-values	17.15**	9.81**	1.32	13.05**	7.27**	10.26**	10.40**	12.42**

** P < .01
*P < .05

educational neglect and neglect as a whole than children having normal health status and excellent health status respectively.

(b) Family Characteristics and Child Maltreatment (Abuse and Neglect)

Present results revealed that the magnitude of child maltreatment varied as a function of family structure, i.e., type of family, family size (number of family members), size of house, and family environment.

1. *Type of Family:* Table 3.4 indicates that children belonging to joint and nuclear families varied significantly on various forms of abuse and neglect. Significant 'F' values suggest that physical abuse, psychological abuse and abuse as a whole, physical neglect, emotional neglect, educational neglect and neglect as a whole were found significantly higher in nuclear families than that of Joint families. It appears that children of nuclear families were found at high risk of abuses and neglects.
2. *Family Size:* The size of family was also identified as one of the predictors of child abuse and neglect. Table 3.5, denotes that children of larger families (>6 members) experienced greater level of physical, psychological and sexual abuses as well as physical, emotional and educational neglects than that of smaller families.
3. *Size of House:* ANOVA results indicated that the size of house exercised impact on child abuse and neglect. Table 3.6 shows that children residing in small houses reported higher degree of physical abuse, psychological abuse, sexual abuse, abuse as a whole, emotional neglect, physical neglect, educational neglect and neglect as a whole than those of average and big size houses. It appeared that high density (crowded physical environment and improper personal space in the house) might have caused high incidences of child abuse and neglect.
4. *The Role of Family Environment in Child Abuse and Neglect:* It is apparent from the results (Table 3.7) that the family environment of abused/neglected and non-abused/non-neglected children, differed significantly. Abused children reported poor cohesion in the family than non-abused

Table 3.4 : Shows Means, SDs and F-values of child abuse and neglect of children in nuclear and joint families

Type of Family	*Abuse*				*Neglect*			
	Physical Abuse	*Psychological Abuse*	*Sexual Abuse*	*Abuse as a Whole*	*Physical Neglect*	*Emotional Neglect*	*Educational Neglect*	*Neglect as a whole*
Nuclear								
Mean	42.45	75.39	9.56	127.46	4.04	17.26	4.95	26.27
S.D.	7.14	12.96	2.62	19.46	1.50	3.61	1.78	6.05
Joint								
Mean	40.20	69.06	9.10	118.62	3.71	16.04	4.38	24.12
S.D.	6.50	12.74	2.06	18.03	1.15	3.42	1.46	5.30
F- values	7.93**	18.15**	2.72	16.29**	4.23*	8.91**	4.96**	10.92**

** P < .01
*P < .05

Table 3.5 : Shows Means, SDs and F-values of child abuse and neglect in smaller and larger families

Size of Family	*Abuse*				*Neglect*			
	Physical Abuse	*Psychological Abuse*	*Sexual Abuse*	*Abuse as a Whole*	*Physical Neglect*	*Emotional Neglect*	*Educational Neglect*	*Neglect as a Whole*
Small Family								
Mean	40.43	70.36	9.06	119.87	3.46	16.46	4.23	24.15
S.D.	5.95	11.29	2.12	15.94	0.99	3.33	1.21	4.50
Large Family								
Mean	42.77	76.17	9.78	128.95	4.19	17.36	5.09	26.65
S.D.	7.38	13.04	2.77	19.76	1.59	3.69	1.84	6.34
F-values	4.29**	9.01**	4.19*	9.88**	7.52**	3.76*	7.83**	6.57**

** P < .01
*P < .05

Table 3.6 : Shows Means, SDs and F-values of child abuse and neglect in accordance with the size of house

Size of house		*Abuse*				*Neglect*			
		Physical Abuse	*Psychological Abuse*	*Sexual Abuse*	*Abuse as a whole*	*Physical Neglect*	*Emotional Neglect*	*Educational Neglect*	*Neglect as a whole*
Small									
	Mean	45.29	78.44	10.03	133.72	4.61	17.76	5.41	27.81
	S.D.	8.02	14.58	3.32	21.99	1.69	3.88	1.99	6.75
Normal									
	Mean	41.49	74.73	9.42	125.96	3.82	17.41	4.89	25.12
	S.D.	6.51	11.55	2.16	16.67	1.26	3.30	1.55	5.18
Big									
	Mean	38.67	66.50	8.81	113.97	3.46	15.25	3.98	22.69
	S.D.	4.88	11.19	1.62	15.09	0.99	3.19	1.24	4.74
F-values		25.24**	24.61**	6.37**	30.79**	19.19**	16.15**	20.46**	22.23**

** $P < .01$
* $P < .05$

Table 3.7 : Shows Means, SDs and 't'-values of family environment, reported by abused/non-abused and neglected/non-neglected children

Group of children		*Dimensions of Family Environment*									
		Cohesion	*Expression*	*Conflict*	*Independence*	*Achievement Orientation*	*Intellectual Cultural Orientation*	*Active Recreational Orientation*	*Moral Religious Emphasis*	*Organization*	*Control*
Abused											
	Mean	20.33	17.03	14.56	20.24	19.54	16.41	13.87	20.03	19.45	18.74
	S.D.	5.01	4.44	4.38	4.86	4.26	4.39	4.14	4.66	4.65	4.31
Non abused											
	Mean	22.07	17.70	12.93	21.43	20.22	17.58	14.27	20.33	19.77	19.22
	S.D.	4.89	3.80	4.15	4.86	4.19	4.09	4.37	5.09	4.66	4.03
	't'-values	2.96**	1.27	3.15**	1.94*	1.11	2.46*	.66	.79	.71	1.30
Neglected											
	Mean	19.75	17.05	15.07	19.13	19.09	16.04	13.78	19.49	18.52	18.53
	S.D.	4.67	4.49	4.28	4.89	4.16	4.70	4.07	4.32	4.66	4.12
Non Neglected											
	Mean	22.06	17.46	12.67	21.01	20.50	17.61	14.35	20.34	19.67	19.17
	S.D.	5.00	3.64	3.79	4.87	3.97	3.86	4.33	5.50	4.40	3.93
	't'-values	3.26**	1.28	4.49**	3.31**	1.52	3.45**	.39	.75	.04	.57

** P < .01,

* P< .05

children. Similarly, on intellectual cultural orientation, abused children scored less than non-abused children. However, abused children perceived high level of conflict in their family environment than that of non-abused children.

Similarly, neglected children perceived their family environment with less cohesion, independence, low achievement orientation and poor intellectual-cultural orientation; however, they reported higher level of conflict in the family. Findings, thus, indicated that family environment also exercised critical role in the origin of child abuse and neglect.

(iii) Influence of Parental Characteristics on Child Abuse and Neglect

Certain parental characteristics, i.e. parental age, gender, marital status, education and health status, influenced significantly to child abuse and neglect. Findings are interpreted below:

1. *Parent's Age:* The ANOVA results revealed that parents' age significantly influenced the incidences of child physical abuse [F (4,330) = 5.75, P<.01], sexual abuse [F (4,330) = 2.85, P<.05] and physical neglect [F (4,330) = 2.54, P<.01]. High rates of physical abuse and neglect were done by Younger parents (below 30 years) [M (phy. abu) = 44.61 and M (phy. neg.) =4.22], whereas, minimum abusive treatments were done by Middle age parents [M (phy. abu.) =39.23 & M (phy. neg.) =3.62]. Contrary to this, reverse pattern was found in case of sexual abuse [M =8.61 *vs.* M = 10.09].
2. *Parent's Gender:* It is also apparent from the result that more negligent treatments were given by mothers than fathers. Significant 'F' values on physical neglect [F (1,333) = 19.09, P<.01] and educational neglect [F (1,333) = 7.74, P< .01] indicated that mothers did more physical neglect (M= 4.27) and educational neglect (M= 5.07) than fathers {M (phy. abu.) = 3.62, M (phy. neg.) = 4.27].
3. *Parent's Education:* The levels of abuse and neglect differed significantly across education level of parents. Table 3.8 indicates that maximum abusive and neglectful treatments were done by illiterate/low educated (0-5 class) parents. However, graduate or post-graduate parents were found less abusive and neglecting.

Table 3.8 : Shows Means, SDs and F-values of child abuse and neglect by parents of different education level

Parent's Education	*Abuse*				*Neglect*			
	Physical Abuse	*Psychological Abuse*	*Sexual Abuse*	*Abuse as a whole*	*Physical Neglect*	*Emotional Neglect*	*Educational Neglect*	*Neglect as a whole*
0-05th Class								
Mean	44.33	77.16	9.60	131.37	4.57	18.14	5.64	28.38
S.D.	7.97	14.28	2.65	21.31	1.67	3.62	1.95	6.30
06-08th Class								
Mean	40.13	70.46	9.21	119.37	3.67	15.88	4.38	23.92
S.D.	4.76	13.08	1.41	16.84	0.92	3.69	1.44	5.01
10-12th Class								
Mean	39.92	70.79	9.27	120.12	3.71	16.02	4.23	23.96
S.D.	6.36	11.27	2.33	17.03	1.19	2.95	1.21	4.76
Graduate and above								
Mean	39.88	70.52	9.28	119.71	3.27	15.96	4.06	23.29
S.D.	5.20	12.03	2.47	16.23	0.71	3.56	1.08	4.68
F- values	11.61**	6.93**	0.50	10.25**	20.05**	10.59**	24.90**	20.63**

** P < .01,
* P < .05

4. *Parent's Health Status* : Results (Table 3.9) denote that parents of different health status differed significantly on physical abuse, psychological abuse, and abuse as a whole, emotional neglect, physical neglect, educational neglect and neglect as a whole. As Table 3.9, indicates, that handicapped/ disabled parents were found highly abusive and neglectful than poor health, normal health and excellent health status parents.

Correlational Analysis

In order to determine the linkages of demographic characteristics of child, family and parents with child maltreatment (abuse and neglect), correlations were computed. Results are displayed in Tables 3.10, (a) & (b).

Relationship between Various Demographic Factors and Child Maltreatment (Abuse and Neglect):

1. *Child Characteristics and Child Abuse/Neglect*

(i) *Child's Age:* Correlations displayed in table 3.10 (a), denote a negative relationship between age and child physical abuse, which suggests that the levels of abuse and neglect decreased with increasing the age of children. Contrary to this, a positive relationship between age and sexual abuse was identified which indicates that with increasing the age, the incidences of sexual abuse increased.

(ii) *Gender:* Correlation results [Table 3.10(a)], further, revealed that gender was found to be positively related with physical neglect, educational neglect and neglect as a whole. Results, thus, indicated that girls as compared to boys were found more physically and educationally neglected than boys.

(iii) *Education:* Results [Table 3.10(a)] revealed that education was found negatively related with child physical abuse, psychological abuse, sexual abuse and abuse as a whole. Similarly, educational neglect and neglect as a whole were also found inversely related with child education. However, physical neglect and emotional neglect were found positively related with education.

Table 3.9 : Shows Means, SDs and F-values of child abuse and neglect by parents of different health status

Parent's Health Status	*Abuse*				*Neglect*			
	Physical Abuse	*Psychological Abuse*	*Sexual Abuse*	*Abuse as a whole*	*Physical Neglect*	*Emotional Neglect*	*Educational Neglect*	*Neglect as a whole*
Handicapped								
Mean	49.00	86.67	11.33	147.00	5.00	21.67	5.67	32.33
S.D.	7.00	13.61	5.77	17.35	1.73	2.52	1.53	2.89
Weak								
Mean	44.39	80.39	10.22	135.39	4.83	17.72	5.78	28.33
S.D.	7.16	18.35	3.44	25.01	1.54	3.66	2.34	6.41
Normal Health								
Mean	41.66	73.09	9.33	124.19	3.90	16.87	4.75	25.53
S.D.	6.97	12.49	2.37	18.68	1.39	3.54	1.64	5.78
Excellent Health								
Mean	36.09	61.91	9.64	107.64	3.00	13.27	3.18	19.45
S.D.	3.51	15.13	1.36	15.95	0.00	2.72	0.41	2.95
F-values	4.46**	5.73**	1.43	6.31**	4.92**	6.05**	5.88**	6.95**

** $P < .01$
*$P < .05$

Table 3.10(a): Shows the relationships of various demographic variables with child abuse and neglect

Demographic variables	*Abuse*				*Neglect*			
	Physical Abuse	*Psychological Abuse*	*Sexual Abuse*	*Abuse as a whole*	*Physical Neglect*	*Emotional Neglect*	*Educational Neglect*	*Neglect as a whole*
Child Health Status	-.274**	-.251**	-.046	-.280**	-.183**	-.265**	-.251**	-.279**
Type of Family	-.153**	-.227**	-.090	-.216**	-.112*	-.161**	-.162**	-.172**
Size of Family	-.006	-.032	-.027	-.028	0.067	-.041	0.043	.003
Size of House	-.362**	-.349**	-.192**	-.392**	-.313**	-.271**	-.326**	-.327**
Parental Age	-.246**	-.035	.180**	-.076	-.124*	-.036	.036	-.059
Parental Gender	0.096	-.046	-.016	-.008	0.233**	-.016	0.151**	0.090
Parental Education	-.283**	-.222**	-.059	-.265**	-.382**	-.268**	-.408**	-.381**
Parental Health Status	-.189**	-.217**	-.083	-.229**	-.202**	-.201**	-.209**	-.233**

** P< .01
* P< .05

(iv) *Birth Order* : Results [Table 3.10 (a)], further, revealed that birth order was found positively correlated with physical neglect and educational neglect and neglect as a whole. Findings suggested that later born were found more neglected than early born children.

(v) *Child's Health Status* : The health status of children was found negatively related with physical abuse, psychological abuse, and abuse as a whole, physical neglect, emotional neglect, educational neglect and neglect as a whole.

Findings, thus, suggested that demographic characteristics of children, viz. age, gender, education, birth order and health status have clear-cut linkages with child abuse and neglect.

2. *Family Characteristics and Child Maltreatment (Abuse & Neglect)*

(i) *Type of family* : Correlation between type of family and child maltreatment (abuse and neglect), was also calculated. Results [Table 3.10(a)] revealed that type of family was found negatively correlated with physical abuse, psychological abuse, abuse as a whole, physical neglect, emotional neglect, educational neglect and neglect as a whole. Results, thus, suggested that child abuse and neglect were found higher in nuclear families than joint families.

(ii) *Size of the family* : The relationships between number of family members (family size) and various forms of child maltreatment were found non-significant.

(iii) *Size of the house* : Correlation results [Table 3.10 (a)] denoted that size of house was found to be negatively related with physical abuse, psychological abuse, sexual abuse and abuse as a whole. Similarly, it was found to be negatively correlated with physical neglect, emotional neglect, educational neglect and neglect as a whole. Results, thus, suggested that with increasing the size of house, the incidences of abuse and neglect decreased. Since, feeling of crowding stress is found higher in residents of small size houses, as a result, children of small houses are found prone to child maltreatment.

3. *Relationships between Family Environment and Child Maltreatment (Abuse and Neglect)*

Correlations were computed between various dimensions of family environment and child maltreatment (abuse and neglect). Results are displayed in Table 3.10(b) and findings are interpreted below :

(i) Results, displayed in Table 3.10 (b), denote that cohesion was found to be negatively related with child physical abuse, psychological abuse, abuse as a whole, physical neglect, emotional neglect, educational neglect and neglect as a whole.

(ii) Expressiveness, was also found to be negatively correlated with child physical abuse, which suggests that poor expressiveness in the family increases the level of physical abuse.

(iii) The relationship between conflict and child maltreatment was also determined. Interestingly, conflict was found to be positively correlated with all the domains of child abuse and neglect, i.e. physical abuse, psychological abuse, abuse as a whole, physical neglect, emotional neglect, educational neglect and neglect as a whole. Results thus, indicated that conflict in family environment is one of the pervasive factors of abusive treatments.

(iv) The relationship between independence and child maltreatment was also established. Independence was found to be inversely related with child physical abuse, psychological abuse, abuse as a whole, physical neglect, emotional neglect, educational neglect and neglect as a whole. Results, thus, suggested that with increasing, the level of independence, the incidence of abuse and neglect decreased.

(v) Achievement orientation was found negatively related with child physical abuse. It appears that with increasing achievement orientation in the family, the incidences of physical abuse decreased.

(vi) Further, intellectual cultural orientation was found to be negatively related with child physical abuse, psychological abuse, abuse as a whole, physical neglect, educational neglect and neglect as a whole.

(vii) Similarly, active recreational orientation was found to be negatively correlated with psychological abuse, physical neglect, educational neglect and neglect as whole.

(viii) However, low but negative relationship between moral and religious emphasis and child maltreatment was found.

Table 3.10 (b) : Shows the relationship of family environment with child abuse and negleet

Dimensions of family environment	*Abuse*				*Neglect*			
	Physical Abuse	*Psychological Abuse*	*Sexual Abuse*	*Abuse as a whole*	*Physical Neglect*	*Emotional Neglect*	*Educational Neglect*	*Neglect as a whole*
Cohesion	-.20**	-.184**	0.03	-.192**	-.16**	-.17**	-.17**	-.20**
Expressiveness	-.108*	-.103	0.086	-.102	-.06	-.04	-.077	-.073
Conflict	0.19**	0.199**	0.044	0.205**	0.166**	0.202**	0.185**	0.213**
Independence	-.215**	-.128*	-.05	-.172**	-.17**	-.161**	-.197**	-.204**
Achievement Orientation	-.152**	-.065	-.057	-.085	-.095	-.072	-.077	-.076
Intellectual Cultural Orientation	-.178**	-.108*	0.018	-.134*	-.211**	-.096	-.187**	-.160**
Active Recreational Orientation	-.070	-.122*	0.024	-.093	-.115*	-.065	-.124*	-.109*
Moral Religious Emphasis	-.057	-.033	0.102	-.026	-.077	-.031	-.033	-.048
Organization	-.128*	-.069	0.064	-.077	-.143**	-.038	-.112*	-.087
Control	-.015	-.018	.166**	0.002	0.001	0.035	-.045	-.010

* $P < .05$
** $P < .01$

(ix) Furthermore, organization was found to be negatively correlated with child physical abuse, physical neglect and educational neglect. Result suggests that better organization in family decreases the level of maltreatment.

(x) Contrary to this, control was found to be positively correlated with sexual abuse.

Findings, thus, evinced that positive dimensions of family environment were found negatively related with child maltreatment, whereas negative dimensions, i.e. conflict and control were found to be positively correlated with abuse and neglect.

4. *The Linkage between Parental Characteristics and Child Maltreatment*

Correlations were also calculated between demographic characteristics of parents and child maltreatment. Results are displayed in table 3.10(a). and interpreted below :

(i) *Parent's Age :* Correlation results [Table 3.10(a)], indicated that parents age was found negatively related with physical abuse and physical neglect. Findings thus, suggested that younger parents were found more abusive and neglecting than older parents. However, a positive correlation between age and sexual abuse indicated that with increasing the age of parents, the incidence of child sexual abuse also increased.

(ii) *Parent's Gender :* Parent's gender was found to be positively correlated with physical neglect and educational neglect. Results, thus, suggested that mothers were found more neglecting.

(iii) *Education :* Correlation results [Table 3.10(a)], further revealed that parental education was found negatively related with child physical abuse, psychological abuse and abuse as a whole. Similarly, education was also found to be negatively related with child physical neglect, emotional neglect, educational neglect and neglect as a whole. Findings, thus, suggested that the extent of child maltreatment decreased with increasing the level of education.

(iv) *Health Status* : Correlation results [Table 3.10(a)], further, suggested that health status of parents was found negatively correlated with child physical abuse, psychological abuse, abuse as a whole, child physical neglect, emotional neglect, educational neglect and neglect as a whole.

Findings, thus, indicate that certain demographic characteristics of the children, family and parents, are found exclusively associated with child abuse and neglect.

Stepwise Multiple Regression Analysis (SMRA)

Prediction of Child Maltreatment by Demographic Characteristics

Correlations, reported in earlier section of the result show that demographic characteristics of children, family and parents were found significantly correlated with various forms of Child Abuse and Neglect. Based on the perusal of correlation results, it was considered appropriate to go for stepwise multiple regression analysis (SMRA) to examine the relative contributions of antecedent factors (demographic characteristics) to criterion variables (child abuse and neglect). The findings of these analyses are in order:

(i) The SMRA result denotes that Child Physical Abuse (CPA) was significantly explained by seven factors. The size of house contributed maximum negatively (β =-.36, R^2=.13, df, 1,333, P<.00), followed by Child education (β = -.27, R^2=.20, df, 2,332, P<.00), Independence in Family Environment (β = -.224, R^2 =.23, df, 3,331, P<.00), Child health status (β = -.16, R^2 =.25, df, 4,330, P<.00), Parental age (β = -.158, R^2 =.27, df, 5,329, P<.000), Residential setting (β = -.113, R^2 =.58, df, 6,328, P<.000) and Type of family (β = -.11, R^2 =.29, df, 7,329, P<.000) respectively.

(ii) Regression analysis further indicated that Child Psychological Abuse (C Psy A) was predicted by 5 demographic factors. The size of house contributed maximum negatively (b=-.35, R^2 =.12, df, 1,333, P<.00) followed by Child health status (β =-.168, R^2 =.148, df, 2,332, P<.00), Type of family (β= -.164, R^2 =.17, df, 3,331, P<.00), Cohesion in family (β= -.14, R^2=.19, df ,4,330, P<.00) and Parental Health (β = -.11, R^2 =.20, df, 5,329, P<.00) respectively.

(iii) The Child Sexual abuse (CSA) was explained by two factors. Child age contributed maximum positively (β = .267, R^2 =.07, df 1,333, P<.01) followed by size of house which contributed negatively to child sexual abuse (β = -.20, R^2=.11, df, 2,332, P<.001).

(iv) The Abuse as a Whole was significantly predicted by 8 demographic variables. The size of house contributed maximum negatively (β = -.39, R^2=.154, df, 1,333, P<.00) followed by child health (β = -.186, R^2=.186, df, 2,332, P<.00), Cohesion in family (β =-.157, R^2 =.21, df, 3,331, P<.00), Type of family (β = -.13, R^2=.225, df, 4,330, P<.00), Child's education (β = -.12, R^2=.24, df, 5, 329, P<.00), Child's age (β = -.19, R^2 =.25, df, 6,328, P<.00), Organization in family (β = -.122, R^2=.26, df, 7,327, P<.00) and Independence (β = -.11, R^2 =.275, df, 8,326, P<.00) .

(v) Regression results further evinced that Child Neglect was contributed by various demographic factors. Child Physical Neglect was explained by 6 factors. Parent's education contributed maximum negatively (β =-.38, R^2=.15, df, 2,333, P<.00) followed by child's education (β = -.19, R^2 =.18, df, 1,333, P<.00). However, positive contributions were made by child's age (β = .268, R^2 =.21, df, 2,332, P<.01). Contrary to this, Intellectual-cultural orientation in the family (b = -.15, R^2 =.23, df, 4,330, P<.01) and Size of house (β =-.13, R^2 =.25, df, 5,329, P<.01) predicted negatively but Child's gender contributed positively (β =-.13, R^2 =.27, df, 6,328, P<.01).

(vi) Child Emotional Neglect was predicted by 7 factors. The Size of house contributed maximum negatively (β = -.27, R^2=.07, df, 1,333, P<.00) followed by Child's health (β = -.21, R^2 =.07, df, 2,332, P<.00), Parent's education (β = -.177, R^2 =.14, df, 3,331, P<.00), however, conflict in family contributed positively (β = .13, R^2=.155, df, 4,330, P<.01). Whereas, parent's gender (β = -.13, R^2=.17, df, 5,329, P<.01), Type of family (β = -.12, R^2 =.18, df, 6,328, P<.01) and Parent's health (β = -.11, R^2 =.187, df, 7,327, P<.01) contributed negatively to child emotional neglect.

(vii) Regression result further indicated that Child's Educational Neglect was predicted by nine (9) factors. In this regard, Parents' education contributed maximum negatively (β =-.41, R^2 =.166, df, 6,332, P<.00), followed by size of house (β =-.21, R^2 =.20, df, 2,332, P<.001), Child's education (b=-.15, R^2 =.22, df, 3,331, P<.001), whereas; Child's Birth order (β = .13, R^2 =.24, df, 4,330, P<.00) and Child's Age contributed positively (β = .175, R^2 =.27, df, 6,328, P<.05). Contrary to this, the Educational neglect was explained negatively by

Child's Health (β =-.11, R^2 =.28, df, 7,327, P<.05), Type of family (β =-.113, R^2=.29, df, 8,326, P<.05) and Independence in family (β =-.11, R^2 =.01, df, 9,325, P<.05).

The Neglect as a Whole was explained by 6 factors. Parent's education contributed maximum negatively (β = -.38, R^2 =.15, df, 1,333, P<.00) followed by child's health status (β = -.215, R^2 =.19, df, 2,332, P<.00), Type of family (b = -.166, R^2 =.22, df, 3,331, P<.00), Independence (β = -.15, R^2 =.24, df, 4,330, P<.00) and Child's education (β = -.13, R^2 =.26, df, 5,329, P<.01) respectively. However, Child's age contributed positively (β = .186, R^2 =.27, df, 6,328, P<.01) to Child neglect.

Results, thus, denoted that Abuse and Neglect along with their various forms, were predicted by a sizable number of demographic characteristics of parents, children and their family in different manner. In other words, mumerous contextual and demographic factors play prominent role in the origin of child abuse and neglect.

Section II

Social Status and Occupational Characteristics: As Predictors of Child Abuse and Neglect

Second objective of the study was to find out the social status (caste and socio-economic status) and occupational characteristics of families, in which children are abused and neglected. The socio-economic status, caste and occupational characteristics of abusive and non-abusive families were identified on the basis of personal data sheet (PDS). Data were analyzed in terms of Means, SDs, Analysis of Variance (ANOVA), Correlations and Stepwise Multiple Regression Analyses.

ANOVA Analysis

1. *Social Status and Child Abuse*

(i) *Socio-economic Status* : Children belonging to high, middle and low socio-economic status families reported different patterns and levels of abuses and neglects. Results displayed in Table 3.11 indicates that :

Table 3.11 : Shows Means, SDs and F-values of child abuse and neglect across different socio-economic status of the families

S.E.S.		*Abuse*				*Neglect*			
		Physical Abuse	*Psychological Abuse*	*Sexual Abuse*	*Abuse as a whole*	*Physical Neglect*	*Emotional Neglect*	*Educational Neglect*	*Neglect as a whole*
Low	Mean	44.83	79.05	9.75	133.82	4.54	18.15	5.55	28.26
	S.D.	7.58	12.55	2.59	18.79	1.58	3.33	1.81	5.81
Middle	Mean	41.22	72.81	9.49	123.54	3.84	16.79	4.70	25.34
	S.D.	6.53	12.57	2.76	18.64	1.38	3.71	1.65	5.82
High	Mean	37.82	65.59	8.80	112.33	3.16	15.05	3.70	21.93
	S.D.	4.10	10.82	1.60	13.47	0.43	3.01	0.72	3.73
F-values		32.14**	32.88**	4.19*	40.69**	30.86**	22.29**	39.01**	37.74**

** P < .01

*P < .05

(i) Physical abuse was found maximum in low SES families than Middle and High SES counterparts,

(ii) Psychological abuse also differed significantly across various SES groups. Low SES children were also found more psychologically abused than Middle and High status counterparts respectively,

(iii) The Low SES children were found more sexually abused than Middle and High SES children respectively,

(iv) Abuse as a whole was also identified higher in low SES group than Middle and High SES families subsequently,

(v) Results evinced that low status children were found more physically neglected than Middle and High status children respectively,

(vi) The Emotional neglect also differed across three SES groups of families. The Low SES children were found more neglected than Middle and High SES children subsequently,

(vii) The Educational neglect also differed significantly across different socio-economic status of the families. In low status families' children were found more educationally neglected than those of Middle and High status families respectively,

(viii) Neglect as a whole also varied significantly across different socio-economic status. Low SES children were found more neglected than Middle and High status children subsequently.

To summarize, although incidences of abuse and neglect are found in different strata of society, however, maximum children of low strata are victims of Child Abuse and Neglect in eastern districts of U.P.

(ii) *Caste and Community :* Results (Table 3.12) revealed that caste and community also exercised impact on child abuse and neglect. ANOVA results indicated that physical abuse, psychological abuse, abuse as a whole, physical neglect, emotional neglect, educational neglect and neglect as a whole differed significantly across different groups. Result, thus, evinced that child maltreatments are found maximum in Muslims and SC (Hindu) than OBC and Forward Caste Hindus.

2. *Occupational Characteristics and Child Abuse*

The occupational characteristics of abusive and non-abusive families

Table 3.12 : Shows Means, SDs and F-values of child abuse and neglect in different caste and community

Children's Caste & Community		*Abuse*				*Neglect*			
		Physical Abuse	*Psychological Abuse*	*Sexual Abuse*	*Abuse as a whole*	*Physical Neglect*	*Emotional Neglect*	*Educational Neglect*	*Neglect as a whole*
Scheduled Caste									
	Mean	43.82	77.72	10.33	131.70	4.25	17.65	5.33	27.23
	S.D.	6.34	13.21	2.82	18.61	1.37	3.24	1.69	5.21
O.B.C									
	Mean	42.59	75.69	9.37	128.11	4.25	17.37	5.15	26.80
	S.D.	7.58	13.32	2.26	20.03	1.60	3.53	1.85	6.25
General									
	Mean	39.77	69.55	9.23	118.42	3.47	15.93	4.15	23.57
	S.D.	5.44	12.25	2.47	16.89	0.97	3.55	1.26	4.96
Muslim									
	Mean	45.82	76.55	9.14	131.95	4.64	18.59	5.68	28.91
	S.D.	10.22	13.26	2.38	22.43	1.81	3.50	2.08	6.74
F-values		8.82**	7.75**	2.27	9.98**	10.96**	6.85**	13.56**	12.07**

** P < .01
*P < .05

were also identified (Table 3.13). Although incidences of abuse and neglect were found in all the occupational groups of families but maximum abusive and negligent treatments were found in labor class or cultivators than business class, whereas, minimum abuse and neglect were found in service class families. Contrary to this, majority of non-abusive cases belonged to either business class or service class. The various forms of abuse and neglect varied significantly across different occupational groups. Findings are systematically presented below in brief :

(i) Physical abuse varied significantly across occupational characteristics of families. Maximum abuse was found in labor class than cultivators and business class whereas; minimum abuse was reported in service class.

(ii) Further, psychological abuse differed significantly across four occupational groups. Psychological abuse was reported maximum in labor class than in cultivators and business class respectively and minimum abuse was reported in service class.

(iii) The sexual abuse was also found maximum in labor class than business class, cultivators and service class families subsequently.

(iv) Abuse as a whole, also differed significantly across the different occupational characteristics of the families. Maximum victims of abuse were identified in labor class than those of cultivators, business class and service class respectively.

(v) Physical neglect also differed significantly across four occupational groups. More specifically, higher levels of neglect were found in labor class than cultivators, business class and service class respectively.

(vi) Child emotional neglect was also found maximum in labor class than cultivators and business class, and minimum in service class.

(vii) Result further, indicated that children belonging to labor class were found more educationally neglected than cultivators, business class and service class respectively.

(viii) Maximum victims of neglect were found in labor class than cultivators, business class and service class respectively.

Table 3.13 : Shows Means, SDs and F-values of child abuse and neglect across different occupational characteristics of parents

Parent's Occupational Characteristics		*Abuse*				*Neglect*			
		Physical Abuse	*Psychological Abuse*	*Sexual Abuse*	*Abuse as a whole*	*Physical Neglect*	*Emotional Neglect*	*Educational Neglect*	*Neglect as a whole*
Labour	Mean	47.39	81.49	10.29	138.96	4.84	18.78	5.96	29.57
	S.D.	8.01	12.67	3.55	20.53	1.61	3.39	1.90	6.04
Cultivation	Mean	42.23	75.94	9.34	128.22	3.95	17.45	4.88	26.28
	S.D.	6.61	12.03	1.90	17.04	1.38	3.57	1.66	5.74
Business	Mean	40.38	70.86	9.42	120.66	3.79	16.73	4.66	25.20
	S.D.	6.64	12.26	1.83	17.87	1.29	3.16	1.52	5.24
Service	Mean	39.93	69.76	9.11	118.74	3.65	15.80	4.29	23.77
	S.D.	5.85	13.04	2.50	18.07	1.25	3.56	1.50	5.49
	F-values	16.62**	12.52**	2.84*	16.99**	9.52**	9.95**	13.09**	13.54**

** P < .01
*P < .05

ANOVA results thus, suggested that families, in which children are highly abused and neglected, were either from labor class, or cultivators' families. Contrary to this, in service class abusive and neglectful treatments were found minimum.

Correlational Analysis

In order to establish the relationships of social status (socio-economic status, caste and community) and parental occupations with child abuse and neglect, correlations were computed. Results are displayed in Table 3.14.

1. Social Status and Child Maltreatment (Abuse and Neglect)

(i) Socio-economic status correlation results (Table 3.14) denote that socio-economic status was found inversely related with child physical abuse, psychological abuse, sexual abuse and abuse as a whole. Similarly, socio-economic status was also found to be negatively related with child physical neglect, emotional neglect, educational neglect and neglect as a whole. Findings, thus, suggested that with increasing the social status of family the incidences of abuse and neglect decreased.

(ii) Caste/community : Table 3.14, further, displays that caste/community, was found to be negatively correlated with psychological abuse, sexual abuse, abuse as a whole, physical neglect, educational neglect and neglect as a whole. Results, thus, suggest that the incidence of abuse and neglect are found higher in Muslim/Hindu— low caste families and with increasing the caste status, levels of child maltreatment decrease.

2. Occupational Characteristics and Child Maltreatment

The correlations displayed in Table 3.14, denote that occupational characteristics of parents was found to be negatively related with child physical abuse, child psychological abuse, sexual abuse and abuse as a whole. Similarly, child physical neglect, emotional neglect, educational neglect and neglect as a whole, were also found inversely related with occupational characteristics of the family. Result, thus, suggests that with increasing the occupational status of the parent's or families, the extent of abuse and neglect decrease.

Table 3.14 : Displays relationships of socio-economic status, caste/community and parental occupation with child abuse and neglect

Variable's Name	*Abuse*				*Neglect*			
	Physical Abuse	*Psychological Abuse*	*Sexual Abuse*	*Abuse as a whole*	*Physical Neglect*	*Emotional Neglect*	*Educational Neglect*	*Neglect as a whole*
Socio-economic Status	-.40**	-.406**	-.152**	-.44**	-.40**	-.343**	-.436**	-.425**
Caste and Community	-.104	-.169**	-.117**	-.169**	-.136*	-.097	-.163**	-.139**
Parental Occupation	-.325**	-.307**	-.134**	-.349**	-.248**	-.285**	-.305**	-.325**

* P < .05
** P < .01

Stepwise Multiple Regression Analysis (SMRA)

Prediction of Child Abuse and Neglect

Based on the perusal of correlation results, it was considered adequate to go for stepwise multiple regression analysis (SMRA) to explore the relative contributions of Social Status and Occupational Characteristics (Predictors) to Child abuse and neglect (Criterion). The results of these analyses are in order:

> Present result evinced that the Child Physical Abuse (CPA) was explained maximum negatively by socio-economic status (β = -.403, R^2 =.16, df, 1,333, P<.00) followed by Parental occupation (β = -.16, R^2 =.18, df, 2,332, P<.00). Similarly, Child Psychological Abuse (C Psy A) was predicted by SES and occupation. The Socio-economic status contributed maximum negatively (β = -.41, R^2 =.165, df, 1,333, P<.00) followed by Parent's occupation (β = -.14, R^2 =.18, df, 1,332, P<.00). Despite this, Child Sexual Abuse (CSA) was negatively explained by only socio economic status (β = -.15, R^2 =.02, df, 1,331, P<.005). Further, the "Abuse as a Whole" was predicted by SES and Occupation of parents/ family. The SES contributed maximum negatively (β = - .44, R^2 =.20, df, 1,333, P<.00) followed by Parent's occupation (β = -.166, R^2 =.217, df, 2,332, P<.01).
>
> Regression result further revealed that Child Neglect was also predicted by social status and occupation characteristics. The Physical neglect was explained maximum negatively by SES (β = -.40, R^2 =.157, df, 1,333; P<.00) followed by Parent's education (β = -.23, R^2 =.19, df, 2,332, P<.00). Further, Emotional Neglect was also predicted by SES and occupation. The socio-economic status contributed maximum negatively (β = -.34, R^2 =.12, df, 1,333, P<.00) followed by Parent's occupation (β = -.15, R^2 =.134, df, 2,332, P<.01). The Educational Neglect was also explained by SES and Parent's Education. The socio-economic status predicted maximum negatively (β = -.44, R^2 =.19, df, 1,333, P<.00) followed by Parent's education (β =-.24, R^2 =.23, df, 2,332, P<.00) respectively. The result denoted that "Neglect as a whole" was also contributed by SES and Parent's education. The

socio-economic status (SES) explained maximum negatively (β = -.42, R^2 =.18, df, 1,333, P<.00) and Parent's Education was the second contributing factor (β = -.21, R^2 =.21, df, 2,332, P<.00).

It is apparent from the result that low socio-economic status, low occupation level and low education of parents favorably predict the high incidence of child abuse and neglect.

Section III

Child Abuse and Neglect in Relation to Psychological Dispositions of Parents

Third objective of the study was to explore the psychological dispositions of abusive and neglecting parents. The abusive and non-abusive groups of parents were identified on the basis of 'Child Abuse-Neglect Interview Schedule for Parents (CANISP). Parental attitude, anxiety, stress, and coping responses were determined on the basis of appropriate measures. Data obtained on parents were treated statistically in terms of Means, SDs, Analysis of Variance (ANOVA), Correlations and Step wise multiple Regression Analysis SMRA. Results are displayed in Tables (3.15 and 3.16)..

Comparative Analysis

ANOVA Results

(i) ***Parental Attitude***: Table 3.15 indicates that Abusive/ Neglecting and Non-abusive/Non-neglecting parents differed significantly on attitude towards their parents. More specifically, abusive parents reported less favorable attitude towards their parents than non-abusive parents. Similarly, neglecting parents also reported unfavorable attitude (restrictive) towards their parents than non-neglecting parents. It appears that abusive/neglecting parents have received negative/abusive experiences in their childhood than their nonabusive counterparts.

(ii) *Anxiety*: The anxiety level also varied significantly across abusive and non-abusive groups. Abusive parents reported higher level of anxiety than non-abusive parents. Furthermore,

Table 3.15 : Shows Means, SDs and Significant 't'-values of attitude towards parents, stress, anxiety and coping responses of parents

		Abusive	*Non-Abusive*	*'t'- values*	*Neglectful*	*Non-Neglectful*	*'t' - values*
Attitude	Mean	125.18	137.32	4.43**	123.82	138.71	5.51**
Towards Parents	S.D.	22.78	27.36		24.17	25.26	
Stress	Mean	49.75	43.34	4.48**	50.91	42.12	6.32**
	S.D.	12.99	13.14		12.50	12.94	
Anxiety	Mean	93.43	85.49	3.10**	94.99	83.85	4.40**
	S.D.	23.39	23.55		22.74	23.56	
Dimensions of Coping							
Positive Reframing	Mean	11.73	12.09	.85	11.43	12.41	2.27*
(Active style)	S.D.	3.48	4.45		2.47	5.08	
Restraint Coping	Mean	10.56	11.11	2.27*	10.59	11.08	2.03*
(Active style)	S.D.	2.26	2.17		2.29	2.14	
Drug Use	Mean	7.87	6.32	3.63**	7.74	6.47	2.94**
(Maladaptive style)	S.D.	4.31	3.41		4.31	3.47	

** P < .01, * P < .05

the level of anxiety differed significantly across neglecting and non-neglecting group of parents. Neglecting parents were found highly anxious than non-neglecting parents (Table 3.15).

(iii) *Stress*: The level of stress also differed significantly in abusive and non-abusive parents. Abusive parents reported higher level of stress than non-abusive parents. Similarly, the level of stress varied significantly in neglecting and non-neglecting parents. More specifically, neglecting parents expressed highless level of stress than non-neglecting parents.

(iv) *Coping with Stress* : Results displayed in Table 3.15, revealed that abusive/neglecting parents differed significantly on coping styles and strategies than their non-abusive/non-neglecting counterparts. The restraint coping (Active) was less exercised by abusive than non-abusive parents. Similarly, neglecting parents as compared to non-neglecting were also found inferior on active related positive reframing, and restraint coping. Contrary to this, abusive parents used drugs (Maladaptive style) as a coping strategy more frequently than non-abusive parents. Similarly, neglecting parents also highly exercised maladaptive style (Drug) than their non-neglecting counterparts.

Finding, thus, suggests that in order to cope with stress, non-abusive/non-neglecting parents exercise more active coping strategies, which have buffering impact to mitigate its consequences whereas, abusive/neglecting parents failed to apply such coping responses properly. However, abusive and neglecting parents exercised maladaptive coping (Drug), which has intensing role in stress-outcome (abuse) relationships.

Correlational Analysis

Correlations were also computed to establish the linkages of child abuse and neglect with parental attitude, stress, anxiety and coping. Results are displayed in Table 3.16, and interpreted below :

(i) Correlation result Table 3.16 evinced that positive attitude towards parents was found to be negatively related with child abuse and neglect. It appears that childhood negative experiences have direct link with abusive/neglecting treatments,

Table 3.16 : Shows relationships of attitude towards parents, stress, anxiety coping responses with child abuse and neglect

Variables Name	*Abuse*	*Neglect*
Attitude Towards Parents	r =-.26**	r =-.28**
Stress	r =.30**	r =.35**
Anxiety	r =.30**	r =.34**
Dimensions of Coping :		
Active coping (active style)	r =-.094	r =-.12*
Planning (active style)	r =-0.15**	r =-.09
Positive Reframing (active style)	r =-.06	r =-.13**
Restraint Coping (active style)	r =-.12*	r =-.11*
Focus on and venting of Emotions (maladaptive)	r =.077	r =.12*
Drug Use (maladaptive)	r =.17**	r =.093

N = 335
**P < .01
*P < .05

(ii) Parental stress was found to be positively related with abuse and neglect,

(iii) Anxiety level was also found to be positively related with abuse and neglect,

(iv) However, varying patterns of relation were found between coping and child abuse/neglect. Active coping was found to be negatively correlated with child neglect. Positive reframing was also found to be negatively correlated with child neglect. Similarly restraint coping was found to be inversely related with abuse and neglect. Planning was found negatively correlated with child abuse. Contrary to this, Maladaptive related focus on and venting of emotions was found to be positively related with child neglect. Similarly, "drug" was also found to be positively related with child abuse.

Results, thus, indicated that active coping responses exercised buffering role (positive) to minimize incidence of child abuse and neglect, whereas, maladaptive coping played intensing role in the process of child maltreatment.

Stepwise Multiple Regression Analysis (SMRA)

Prediction of Child Abuse and Neglect : On the basis of correlation results; it was viewed essential to calculate stepwise multiple regression analysis (SMRA) to examine the relative contributions of certain psychological characteristics of parents to child abuse and neglect (criterion variables).

The child abuse (as a whole) was predicted significantly by two factors. Stress in parents was found the strongest factor, which explained positively (β = .32, R^2=.10, df, 1,333, P<.00) followed by attitude, which contributed negatively (β = -.16, R^2 =.124, df, 2,332, P<.01).

Regression result, further evinced that child neglect (as a whole) was explained by three (3) factors. The attitude predicted maximum negatively (β=-.67, R^2 =.12, df, 2,333, P<.00) followed by Stress, which contributed positively (β=-.42, R2=.16, df, 2,332, P<.00) and Adaptive coping (support of competing activity) style (β=-.197, R^2 =.17, df, 3,331, P<.05) that predicted negatively to child abuse and neglect. Present results thus, suggest that confavarable parental attitude, high stress and anxiety and maladabtive coping style in parentss lead to high incidence of child abuse and neglect.

Section IV

The Consequences of Child Abuse

The fourth objective of study was to measure the outcomes of child abuse on cognitive, motivational and behavioural functionings and personality pattern of children. Data obtained on various measures were scored and treated statistically in terms of Means, SDs, Analysis of Variance (ANOVA), Correlations and Stepwise Multiple Regression Analysis (SMRA). Graphic representation of data was also done. Obtained results are displayed and interpreted into four sub sections.

(A) Cognitive Processes as a Function of Child Abuse, Environmental Setting, Gender, Age and SES of Children

ANOVA Analysis

Data, obtained on various Cognitive Measures, viz. Embedded Figure Test (perceptual identification), Tasks of Verbal Learning and Memory

Table 17(a) : Displays Means, SDs and significant 'F' values of Prerceptual Identification, Verbal Organization (Clustering), Memory Span and Intelligence as a function of level of abuse, environmental setting, and gender

Variables	Setting	*Abused*				*Non-Abused*				*F-Value*
		Rural		*Urban*		*Rural*		Urban		F_A = *Child Abuse*
		Boys	*Girls*	*Boys*	*Girls*	*Boys*	*Girls*	*Boys*	*Girls*	F_B = Environmental F_C = Gender
Perceptual Identification										$F_A = 25.01$**
										$F_B = 10.04$**
	Mean	28.06	25.41	29.52	30.38	31.17	30.91	32.63	30.53	$F_{AXB} = 5.10$**
	S.D.	6.39	6.21	6.16	5.23	4.85	4.63	4.34	4.21	$F_{AXBXC} = 5.08$*
Clustering										
List-A	Mean	9.26	9.69	10.82	10.50	11.07	10.18	12.21	11.98	$F_A = 15.50$**
	S.D.	3.13	3.43	2.68	2.61	3.39	3.41	2.51	2.66	$F_B = 16.35$**
List-B	Mean	8.48	8.85	10.00	9.83	9.30	8.67	11.85	11.53	$F_A = 8.83$**
	S.D.	3.36	3.30	3.73	2.93	3.76	3.19	2.52	2.69	$F_B = 31.57$**
Memory Span										$F_A = 43.20$**
	Mean	3.44	2.85	3.91	3.79	4.47	3.85	5.44	4.43	$F_B = 21.45$**
	S.D.	1.43	1.44	1.36	1.62	1.32	1.46	1.20	1.62	$F_C = 13.46$**
Intelligence										$F_A = 63.91$**
	Mean	64.98	66.21	65.52	69.76	75.23	72.64	84.77	84.31	$F_B = 25.93$**
	S.D.	10.21	7.29	10.99	10.78	15.89	10.14	14.26	19.59	$F_{AXB} = 4.98$*

* $P < .05$
** $P < .01$

Span, Tests of Creative Thinking and Culture Fair Intelligence Test, were scored according to the defined rules given in manuals. Several statistical analyses were done and results are displayed in Tables [3.17(a) and (b) and 3.18(a) and (b)] and figures [3.1, 3.2, 3.3 (a) and (b) 3.4, 3.5, 3.6 and 3.7] respectively.

(i) *Perceptual Identification :* Scored data were treated statistically in terms of Means, SDs and ANOVA and are displayed in Table 3.17(a). Results indicated a significant main effect for Child Abuse, which revealed that abused children performed poor on perceptual identification (M =28.34) than non-abused children (M = 31.31). Perceptual identification also varied significantly in rural-urban settings. Rural children performed poor on perceptual identification (M =28.89) than urban children (M = 30.77). However, significant Child Abuse × Environmental setting interaction effect (Interaction graph-3.1), revealed that in case of non abused group, very little differences on perceptual identification were found between rural (M = 31.04) and urban children (M = 31.59) however, abused children of urban setting made more correct identifications (M = 29.95) than rural children (M = 26.74). Further Child Abuse × Environmental setting × Gender interaction effect (Figure 3.2) indicated that abused boys, belonging to both, rural and urban settings performed inferior as compared to non-abused boys but different pattern was identified in case of girls. Urban girls of abused and non-abused groups did not differ significantly but in case of rural sample perceptual identification was very poor in abused girls as compared to non-abused counterparts.

In a further analysis [Table 3.17(b)], significant main effect for socio-economic status denoted that low SES children made poor perceptual identification (M = 28.24) than middle (M = 30.28) and high (M = 31.80) SES children subsequently. Perceptual identification also varied significantly across age. Children performed less correctly (M = 26.94) than pre-adolescents (M = 30.87) and adolescents (M = 32.512) respectively.

(ii) *Impact on Verbal Learning and Memory :* The ANOVA result [Table 3. 17(a)] evinced the influence of child abuse and setting on verbal organization. Significant main effect for

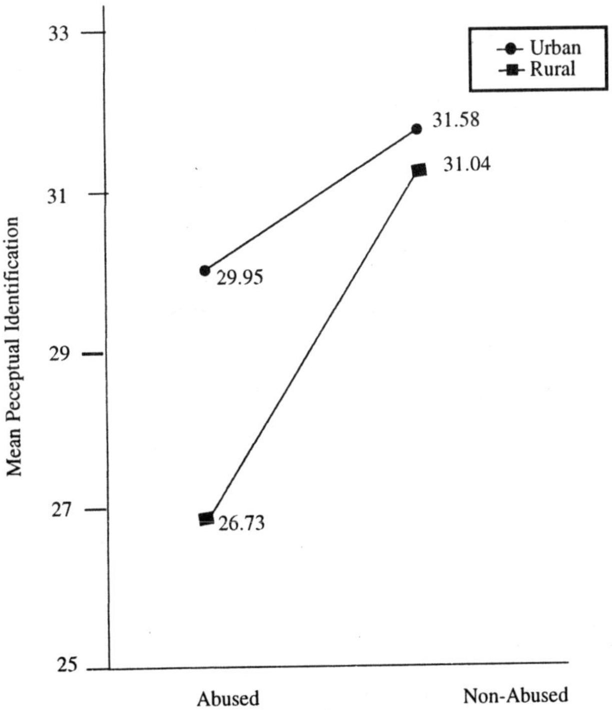

Fig. 3.1 : Perceptual identification as a function of interaction of the level of abuse and environmental residential setting

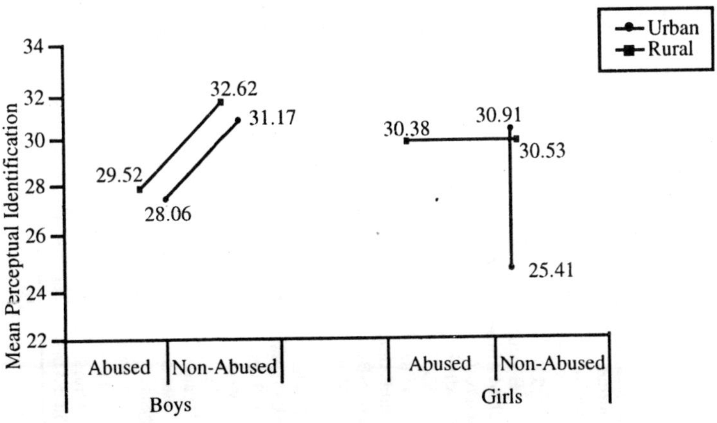

Fig. 3.2 : Perceptual identification as a function of three way interaction of the level of abuse, environmental setting and gender.

Table 3.17(b) : Displays Means, SDs and Significant 'F'-values of Perceptual Identification, Verbal Organization (Clustering), Memory Span and Intelligence as functions of level of abuse, SES and age

Variables	*Abused*									*Non-Abused*								
	Low			*Middle*			*High*			*Low*			*Middle*			*High*		
	6-10 Yrs.	*11-14 Yrs.*	*15-18 Yrs.*	*6-10 Yrs.*	*11-14 Yrs.*	*15-18 Yrs.*	*6-10 Yrs.*	*11-14 Yrs.*	*15-18 Yrs.*	*6-10 Yrs.*	*11-14 Yrs.*	*15-18 Yrs.*	*6-10 Yrs.*	*11-14 Yrs.*	*15-18 Yrs.*	*6-10 Yrs.*	*11-14 Yrs.*	*15-18 Yrs.*
Perceptual Identification																		
Mean	24.03	27.44	30.20	26.94	29.52	32.25	26.71	32.60	33.71	26.40	30.21	31.15	28.89	31.47	32.60	28.64	34.00	35.16
S.D.	5.83	5.33	4.55	7.12	6.36	5.25	7.91	4.40	2.43	5.10	5.99	2.94	5.03	2.74	3.79	3.36	2.34	2.54
F-values																		
FA (Child Abuse) = 8.12, FB (SES) = 11.66**& FC (Age) = 29.83****																		
Clustering-A																		
Mean	8.18	9.18	10.92	10.11	9.48	12.31	10.14	12.20	13.43	9.00	9.07	12.23	10.21	12.29	12.75	9.91	12.85	13.28
S.D.	2.91	1.95	2.68	2.59	3.85	2.06	2.73	2.35	1.51	3.37	3.43	2.86	3.72	2.39	1.94	2.58	2.06	1.54
F-values																		
FA (Child Abuse) = 3.77*, FB (SES) = 15.60& FC (Age) = 25.51****																		
Clustering-B																		
Mean	7.52	8.12	10.16	9.44	8.84	11.69	9.43	11.50	12.14	8.10	8.71	10.15	9.58	11.71	11.50	9.73	11.95	12.00
S.D.	3.48	2.75	3.17	3.26	3.10	2.91	4.30	2.51	2.50	3.50	4.30	1.91	3.93	2.42	2.76	3.55	2.50	2.16
F-values																		
FB= SES (Low, Middle and High)=13.73 and FC= Age (6-10Yrs, 11-14 Yrs and 15-18 Yrs)=11.99****																		
Memory Span																		
Mean	2.30	3.06	3.72	3.22	4.16	4.37	4.14	4.50	5.00	3.50	3.64	5.23	3.68	4.59	4.95	4.18	5.40	5.52
S.D.	1.21	1.40	1.31	1.17	1.25	1.20	1.57	1.35	1.91	0.97	1.00	1.36	1.91	1.54	1.43	1.10	1.19	1.29
F-values																		
FA= (Child Abuse) = 18.24, FB (SES) = 17.34**, FC (Age) = 20.21****																		
Intelligence																		
Mean	62.79	63.21	67.56	69.39	69.64	69.50	75.86	69.50	84.29	67.20	70.93	74.31	72.21	86.29	79.20	80.14	90.45	89.12
S.D.	4.82	5.29	8.24	6.87	11.68	13.84	8.03	13.35	15.95	4.73	9.06	18.78	8.28	19.42	14.96	7.78	19.06	18.88
F-values																		
FA (Child Abuse) = 34.25, FB (SES) = 26.96** , FC (Age) = 5.36** & F AXC = 5.34****																		

**P< .01,

*P<.05

Child Abuse [F (1,327) =15.50, P < .01] revealed that abused children showed poor clustering in verbal organization (M= 10.07) than non-abused children (M =11.36). Significant main effect for Environmental Setting [F (1,327) = 16.35, P < .01] also indicated that rural children showed poor verbal organization (clustering) (M =10.05) than urban children (M = 11.38). In addition, verbal clustering varied significantly across different SES groups and Age [Table 3.17(b)]. Results suggested that clustering was superior in high SES children (M = 12.49) than middle (M = 11.19) and low SES (M = 9.76) counterparts. The clustering level in list-B also varied significantly across Child Abuse and Environmental Setting. Abused children made high clustering responses (M = 9.29) than non-abused children (M =10.34). Similarly, rural children showed poor clustering (M = 8.82) than urban children (M = 10.80). Furthermore, Table (3.17, b) evinces that clustering of list- B varied significantly across different SES groups and age. High SES children displayed superior clustering (M =11.12) than middle (M =10.46) and low SES children (M = 8.83). Similarly, adolescents did better clustering (M = 11.27) than pre adolescents (M = 10.14) and children (M =8.97).

(iii) *Memory Span* : Result, further, [Table 3.17(a)] indicated the varying patterns of memory span across different groups of children. Mean scores of memory differed significantly across child abuse, environmental setting and gender. Significant main effect for Child Abuse revealed that abused children showed limited memory span (M= 3.5) than non-abused children (M = 4.55). Similarly, significant main effect for Environmental Setting indicated that rural children expressed lower span of memory (M = 3.65) than urban children (M= 4.39). The level of memory span also differed significantly in boys and girls. Boys showed superior memory span (M = 4.31) than girls (M = 3.73). Memory span also differed significantly across SES and Age level. Results revealed that high SES children showed better memory span (M = 4.79) than middle (M= 4.23) and low SES children (M = 3.57). Similarly adolescents expressed better memory span (M = 4.80) than pre-adolescents (M = 4.39) and children (M = 3.49).

(iv) *Impact on Creative Potential of Children:* ANOVA results

Table 3.18(a) : Displays Means, SDs and Singnificant 'F'- values of Creative Thinking as a function of level of abuse environmental setting and gender

Variables of Creative Thinking		Abused				Non-Abused				F-Value FA = *Child Abuse* FB = *Residential Setting* FC = *Gender*
		Rural		*Urban*		*Rural*		Urban		
		Boys	*Girls*	*Boys*	*Girls*	*Boys*	*Girls*	*Boys*	*Girls*	
Verbal :										
Fluency	Mean	28.52	24.79	29.09	26.05	29.00	31.42	37.42	37.06	FA= 19.76**
	S.D.	9.95	9.71	11.56	11.67	13.71	17.36	13.61	17.78	FB= 7.12** FA X B= 4.23*
Flexibility	Mean	18.38	15.33	18.80	16.79	18.73	17.76	23.29	23.02	FA= 22.48** FB= 16.83**
	S.D.	5.56	5.19	6.80	6.13	7.35	6.02	6.05	7.85	FC= 4.89* FA X B= 7.79**
Originality	Mean	4.26	3.10	9.64	7.12	9.23	5.58	14.92	12.96	FA= 16.99** FB= 24.86**
	S.D.	4.06	1.15	5.32	4.45	7.81	3.25	12.13	10.25	FC= 4.25*
Non-Verbal :										
Elaboration(N) (Picture)	Mean	21.84	21.90	35.43	34.95	39.87	31.39	44.12	41.76	FA= 32.14**
	S.D.	11.49	13.93	16.93	20.27	23.99	14.47	18.47	16.51	FB= 29.57**
Elaboration(V) (Title)	Mean	12.04	8.69	14.84	i0.21	17.10	11.39	21.06	19.29	FA= 21.24** FB= 10.46**
	S.D.	10.10	6.71	11.32	12.48	11.63	9.71	12.61	12.35	FC= 9.55**

(Contd..)

Table 3.18(a) *(Contd..)*

Originality (N)										FA= 21.65**
(Picture)	Mean	13.18	13.46	24.23	17.62	23.07	16.21	26.88	25.41	FB=32.36**
	S.D.	7.85	10.95	13.78	12.71	11.79	9.46	10.48	11.54	FC= 8.72**
										FA X B X C= 6.13**
Originality (V)										
(Title)	Mean	1.06	0.92	4.39	3.83	5.20	1.61	6.60	6.84	FA= 12.46**
	S.D	1.01	0.21	3.25	1.32	3.31	0.09	4.26	5.64	FB= 20.46**
Flexibility										FA= 26.40**
	Mean	5.40	4.85	7.00	5.21	7.83	6.64	8.54	8.06	FB=6.00**
	S.D.	4.29	3.80	3.48	3.07	4.56	4.37	3.51	3.26	FC= 5.74**

* $P < .05$
** $P < .01$

evinced that level of creative potentials (on various dimensions) varied across child abuse, rural-urban setting, SES and age; however, gender difference was significant only on a few domains of creativity. Results revealed that;

1. *Verbal Fluency* was influenced by child abuse, environmental setting, SES and age. Significant main effect for Child Abuse revealed that abused children achieved far less on verbal fluency (M = 27.11) than that of non-abused children (M = 33.73). The impact of Environmental Setting was also significant which evinced that rural children scored very poor

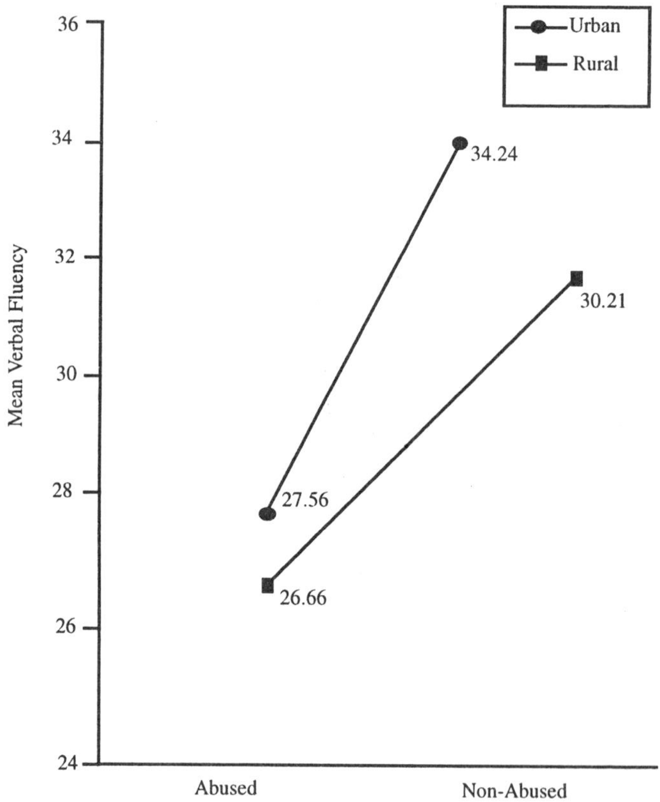

Fig. 3.3 (a) : Verbal fluency as a function of interaction of the level of abuse and environmental setting

on verbal fluency (M = 28.43) than that of urban children (M = 32.40). Child Abuse × Environmental Setting interaction effect was also found to be significant [Table 3.18(a)]. Interaction graph [Figure 3.3(a)] indicates that this ability was found inferior in abused children, belonging to both rural and urban settings, however, in case of non-abused group, urban children performed significantly better than rural counterparts. Further, significant main effect for SES indicated that verbal fluency was inferior in low SES children (M = 25.78) than middle (M = 31.30) and high SES (M = 36.63) children. Age influence was also found significant, which denoted that adolescents expressed superior level of verbal fluency (M = 36.69), than preadolescents (M = 28.56) and children (M = 28.47). Further SES × Age interaction effect [Fig.3.3(b)] denoted that children of three age groups did not differ in low SES group but in middle and high SES groups, they displayed very poor fluency than preadolescents and adolescents respectively.

2. *Verbal Flexibility* : Significant main effect for Child Abuse

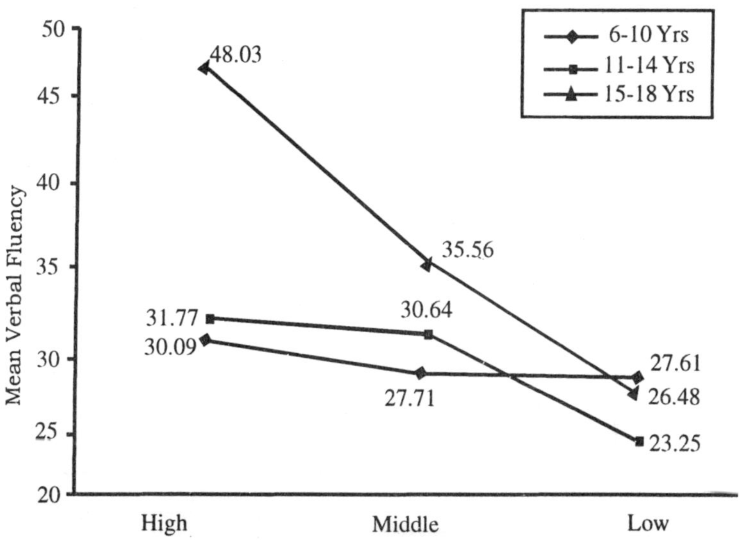

Fig. 3.3 (b) : Verbal fluency as a function of interaction of the level of socio-economic status and age

suggested that abused children scored poor on verbal flexibility (M =17.32) than non-abused chiidren (M = 10.67). Significant main effect for setting revealed that urban children scored higher on verbal flexibility (M = 20.47) than their rural counterparts (M = 17.55). Gender effect was also found significant which denoted that girls expressed low level of flexibility (M = 18.22) than the boys (M = 19.80). Child Abuse × Setting interaction effect was also found significant. Interaction graph (Fig.3.4) denotes that abused children of both settings (rural and urban) expressed poor verbal

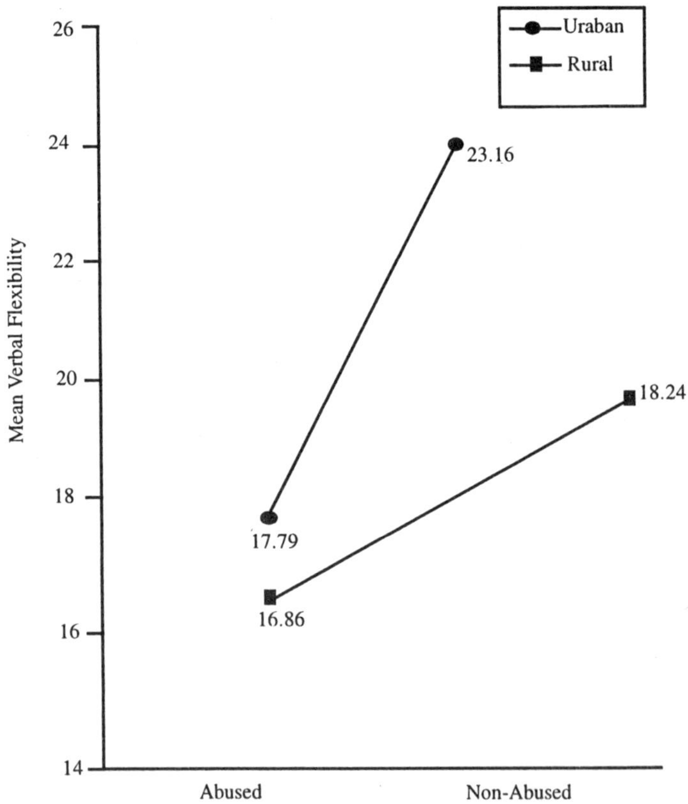

Fig. 3.4 : Verbal flexibility as a function of interaction of the level of abuse and environmental setting

Table 3.18 (b) : Displays Means, SDs and Significant 'F'- values of Creative Thinking as a function of level of abuse, SES and age

Variables		*Abused*									*Non-Abused*								
		Low			*Middle*			*High*			*Low*			*Middle*			*High*		
		6-10 Yrs.	*11-14 Yrs.*	*15-18 Yrs.*	*6-10 Yrs.*	*11-14 Yrs.*	*15-18 Yrs.*	*6-10 Yrs.*	*11-14 Yrs.*	*15-18 Yrs.*	*6-10 Yrs.*	*11-14 Yrs.*	*15-18 Yrs.*	*6-10 Yrs.*	*11-14 Yrs.*	*15-18 Yrs.*	*6-10 Yrs.*	*11-14 Yrs.*	*15-18 Yrs*
Verbal:																			
Fluency	Mean	25.12	23.00	25.96	26.06	28.64	32.56	30.14	27.20	45.43	30.10	23.50	27.08	29.37	32.65	38.55	30.05	35.35	50.64
	S.D.	7.94	7.73	10.26	10.59	13.60	9.84	11.96	9.85	9.60	10.64	8.10	10.40	11.53	15.90	17.84	10.30	12.10	20.15
F-values FA (Child Abuse) = 6.60**, FB (SES) = 16.78**& FC (Age) = 13.42**,FBXC = 4.70**																			
Flexibility	Mean	15.73	14.50	17.92	16.28	18.30	21.13	20.14	18.80	24.60	17.40	14.36	18.46	18.95	21.65	22.75	19.50	23.00	28.28
	S.D.	4.63	4.90	6.00	5.83	7.21	4.43	8.32	6.02	4.80	5.70	5.10	4.81	6.92	7.40	5.53	7.02	5.24	6.96
F-values FA (Child Abuse) = 6.91**, FB (SES) = 22.03**& FC (Age) = 13.23**																			
Originality	Mean	3.73	3.24	6.60	3.89	6.84	12..25	11.29	7.40	10.00	6.20	2.71	5.70	8.84	9.88	11.15	8.64	16.15	22.64
	S.D.	2.53	2.61	5.24	2.70	4.01	10.43	9.10	5.33	8.25	5.10	1.20	3.22	5.10	7.10	8.42	7.02	11.50	17.31
F-values FA (Child Abuse) = 6.44**, FB (SES) = 14.60**, FC (Age) = 5.10**and FAXBXC = 2.41*																			
Non-Verbal :																			
Elaboration (N) (Picture)	Mean	17.70	24.91	29.68	26.00	32.96	38.94	33.00	38.40	38.57	29.40	25.71	29.62	34.42	35.82	43.10	37.41	50.40	56.04
	S.D.	10.83	12.20	19.70	12.70	20.61	16.60	14.64	24.94	8.90	12.31	10.52	11.40	20.23	15.11	14.74	13.32	17.61	22.03
F-values FA (Child Abuse) = 12.25**, FB(SES) = 21.44**& FC (Age) = 7.66**																			

(Contd..)

Table 3.18 (b) *(Contd...)*

Elaboration (v) (Title)																			
	Mean	6.21	8.56	12.48	9.50	13.40	18.00	13.86	18.30	20.57	14.00	13.00	15.46	13.74	15.00	21.30	13.32	21.40	26.36
	S.D.	6.00	5.23	10.62	8.86	12.34	11.86	7.71	13.94	9.90	10.64	9.22	10.70	11.54	13.16	10.90	9.86	10.35	14.35
F-values FA (Child Abuse) = 7.71**, FB (SES) = 9.81**& FC (Age) = 9.68**																			
Originality (N) (Picture)																			
	Mean	10.24	13.71	19.04	14.50	19.08	26.56	23.71	20.10	25.71	17.20	15.71	18.31	18.16	21.06	26.65	21.55	33.45	30.24
	S.D.	9.44	9.13	13.37	9.18	11.72	16.53	14.12	11.20	6.31	8.26	6.83	11.44	11.20	9.92	10.13	10.91	10.80	9.31
F-values FA (Child Abuse) = 6.41**, FB (SES) = 18.82**& FC (Age) = 8.77**																			
Originality (V) (Title)																			
	Mean	0.67	0.88	3.28	1.61	4.04	5.44	3.57	5.80	1.30	3.60	1.79	1.69	6.05	4.88	6.45	3.59	6.70	9.52
	S.D.	0.03	0.61	2.70	1.00	2.30	3.82	3.20	3.41	1.20	2.10	0.80	0.38	4.20	3.21	3.23	2.31	5.30	6.24
F-values FA (Child Abuse) = 6.42**, FB (SES) = 8.76**																			
Flexibility																			
	Mean	3.73	5.06	5.84	5.39	5.64	8.00	6.14	6.20	10.57	6.50	6.50	7.92	5.79	7.00	10.25	7.10	8.40	9.71
	S.D.	2.40	3.42	4.50	3.47	3.41	3.83	3.80	3.71	4.43	3.24	4.51	4.51	2.91	3.91	3.61	3.96	3.11	3.41
F-values FA (Child Abuse) = 10.27**, FB (SES) = 7.15**& FC (Age) = 15.10**																			

** P<.01
 * P<.05

flexibility, whereas, in case of non-abused group, urban as compared to rural children showed superior level of verbal flexibility. Further, Table 3.18(b) denotes that SES also exercised influence on flexibility. Low SES children expressed very poor level of flexibility (M = 16.39) than middle (M = 19.84) and high SES children (M = 22.38) respectively. Age difference was also found highly significant which indicated that adolescents expressed very superior level of flexibility (M = 22.17) than preadolescents (M = 18.43) and children (M = 18.0).

3. ANOVA results, further, denoted that child abuse, environmental setting, gender, SES and age exercised significant impact on Verbal Originality in children. Significant main effect for Child Abuse suggested that abused children scored very low on originality (M = 6.03) than non-children scored very low on originality (M = 6.03) than non-abused counterparts (M = 10.67). Further, significant main effect for Environmental Setting indicated that rural children expressed lower level of originality (M = 5.54) than urban children (M = 11.16). Result, further, suggested that girls were found inferior on this ability (M = 7.19) than boys (M = 9.51). Table 3.18(b), denotes that children belonging to different Socio-economic Status (SES) varied significantly on originality. This ability was found very poor in low SES children (M = 4.69) than middle (M = 8.81) and high SES

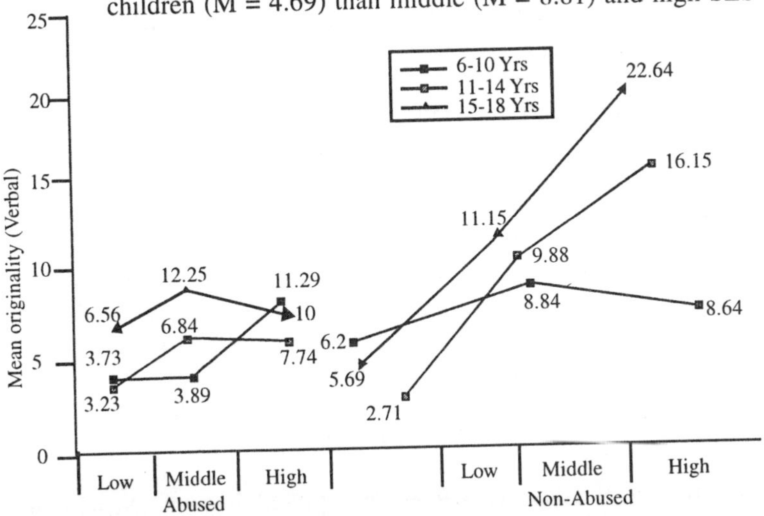

Fig. 3.5 : Originality (verbal) as a function of three-way interaction of level of abuse, socio- economic status and age.

children (M = 12.68). Age differences were also found significant which denoted that adolescents expressed higher level of flexibility (M =11.36) than preadolescents (M = 7.70) and children (M = 7.10). Significant Child Abuse × SES × Age interaction effect (interaction graph 3.5) revealed that originality was found adverse in abused children of low than high and middle SES children but in non-abused group, originality was found quite poor only in low SES group as compared to middle and high SES children.

4. Various dimensions of non-verbal creativity were also influenced by child abuse, setting, SES and age. On Elaboration N (for picture), significant main effect for Child Abuse indicated that abused children showed poor elaboration of pictures (M = 28.53) than non-abused children (M = 39.28). Significant main effect for Environmental Setting also revealed that rural children (M = 28.75) were found inferior than their urban counterparts (M = 39.06). Significant main effect for SES revealed that high SES children (M = 42.30) expressed very high level of elaboration for picture than middle (M = 35.21), and low SES groups (M = 26.17) respectively. Table 3.18(b), further indicated that age differences were found significant. Result, thus, suggested that children showed poor elaboration for picture (M = 29.65) than pre-adolescents (M = 34.70) and adolescents (M = 39.32) respectively.

Similarly, on Elaboration V (for title), significant main effect for Child Abuse indicated that abused children (M = 11.45) were found inferior on elaboration for title (V) than non-abused children (M = 17.21). Furthermore, significant main effect for Setting suggested that urban children were found superior (M =16.35) than rural children (M = 12.31) on this potential. Furthermore, significant main effect for Gender indicated that girls (M = 2.39) were found inferior on elaboration for title than boys (M = 16.26). Table 3.18(b), further, evinced that elaboration of titles differed significantly among high (M =18.97), middle (M =15.15) and low SES children (M = 11.60), [$F(2,317) = 9.81$, $P < .01$]. Age differences were also found significant which denoted that adolescents made very superior level of elaborations of titles (M =19.03) than pre-adolescents (M =14.94) and children (M =11.76) respectively.

5. (i) Creativity related to Originality N (for picture) varied significantly across abuse, setting and gender [Table 3.18(a)]. Significant main effect for Child Abuse evinced that abused children expressed poor level of originality (M =17.12) than non-abused children (M =22.89). Significant main effect for Environmental Setting indicated that rural children (M =16.48) expressed poor level of originality than their urban counterparts (M =23.53). Furthermore, significant main effect for Gender denoted that boys expressed superior level of originality (M =21.24) than girls (M =18.17). Significant Child Abuse × Setting × Gender interaction effect (Fig. 3.6) suggested that abused boys belonging to rural setting expressed poor level of originality in pictures than their urban counterparts but in case of non-abused boys, children of rural and urban setting did not differ significantly. However, reverse pattern was observed in case of girls. Abused girls, of both settings, expressed poor originality but in non-abused group, urban girls expressed very superior level of originality in pictures than rural counterparts. Table 3.18(b), displays that Low SES children showed low level of originality for pictures

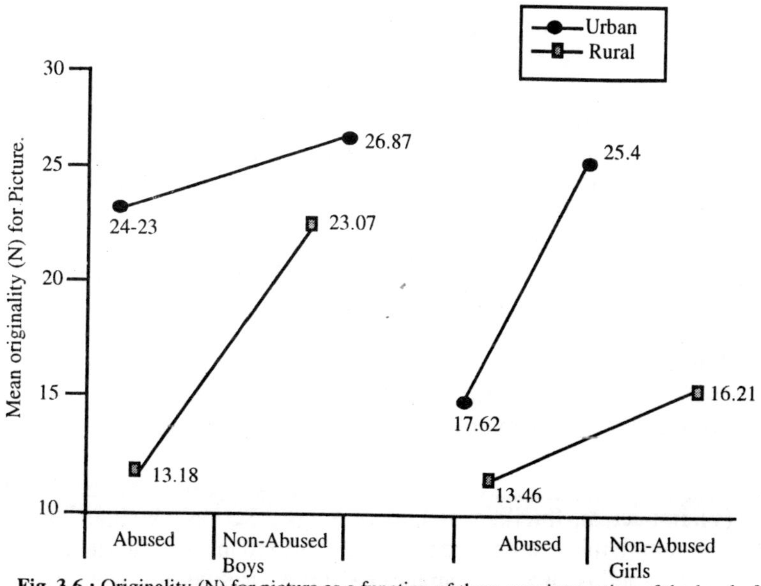

Fig. 3.6 : Originality (N) for picture as a function of three-way interaction of the level of abuse, environmental setting and gender.

(M =15.70) than middle (M =21.02) and high SES (M =25.78) respectively. Significant main effect for Age indicated that adolescents expressed superior level of originality (M =24.40) than pre-adolescents (M =20.52) and children (M =17.56) respectively.

(ii) Further, on Originality 'V' (for title) the effect of Child Abuse was significant. Result, thus, revealed that non-abused children (M =5.06) showed superior level of originality for titles than abused children (M =2.55). Significant main effect for Setting evinced that rural children (M =2.19) expressed low level of originality for title than their urban counterparts (M =5.41). Gender effect was non-significant. Table 3.18(b), further, denotes significant main effect for SES, which suggested that low SES children performed very poor on originality for titles (M = 1.99) than middle (M = 4.75) and high SES groups (M = 5.08) respectively.

6. Likewise, on Non-verbal Flexibility, significant main effect for Child Abuse suggested that abused children (M = 5.61) expressed low level of flexibility than non-abused children (M = 7.77). In addition, significant main effect for Setting revealed that flexibility was found poor in rural children (M = 6.18) than urban children (M = 7.20). Further, girls (M = 6.19) expressed low level of flexibility (non-verbal) than boys (M = 7.19). Result [Table 3.18(b)] evinced that flexibility was inferior in low SES children (M = 5.92) than middle (M = 7.01) and high SES group (M = 8.01) respectively. Significant main effect for Age also denoted that adolescents expressed superior flexibility (M = 8.71) than pre-adolescents (M = 6.47) and children (M = 5.77) respectively.

Findings thus suggested that creative potentials were adversely influenced by child abuse, rural setting, and low SES level of children. Age and gender also exercised significant influence on some dimensions of creativity.

(v) Impact on Intellectual Functionings : Scores obtained on culture fair intelligence test were analyzed in terms of Means, SDs and Analysis of Variance (ANOVA). Results are displayed in Tables 3.17 (a) and (b) and Figures 3.7(a) and (b).

ANOVA results [Table 3.17(a)] indicated that the level of intellect varied across different groups of children. Significant main effect for Child Abuse revealed that abused children scored low on intelligence test (M = 67.66) than non-abused children (M = 79.24), similarly, significant main effect for Environmental Setting denoted that rural children were found intellectually inferior (M = 69.76) than urban children (M = 77.14). Significant child abuse × setting interaction effect [Fig.3.7(a)] denotes that abuse and setting exercised joint impact on intelligence level of children. In case of abused group, very little variation on intellectual functioning was found between rural and urban children, however, a clear difference was identified between rural and

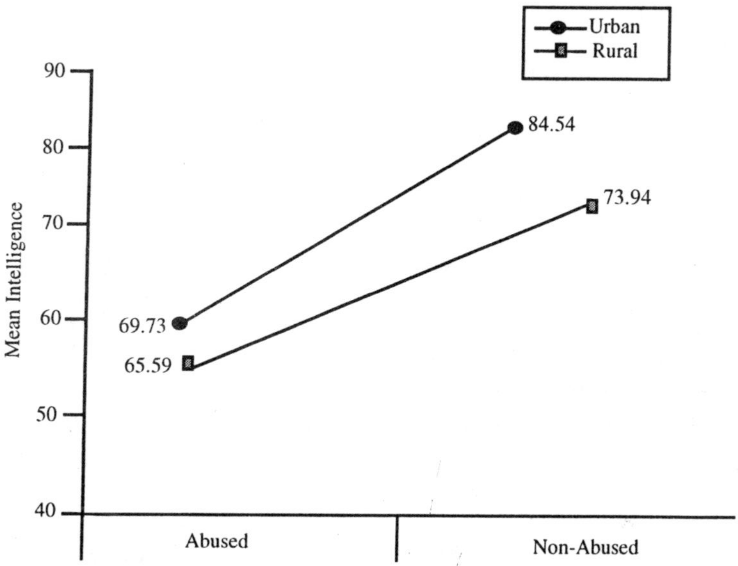

Fig. 3.7 (a) : Intelligence as a function of interaction of abuse and age

urban sample of non-abused group. Non-abused urban children showed superior level of intelligence than rural children.

In a further set of ANOVA analysis [Table 3.17(b)], significant main effect for SES revealed that high SES children performed better on intellectual functioning (M = 81.56) than middle (M = 74.37) and low (M = 67.67) SES children. Further, significant main effect for Age indicated that children scored lower (M = 71.26) than Pre-adolescents (M = 75.00) and Adolescents (M = 77.33) respectively

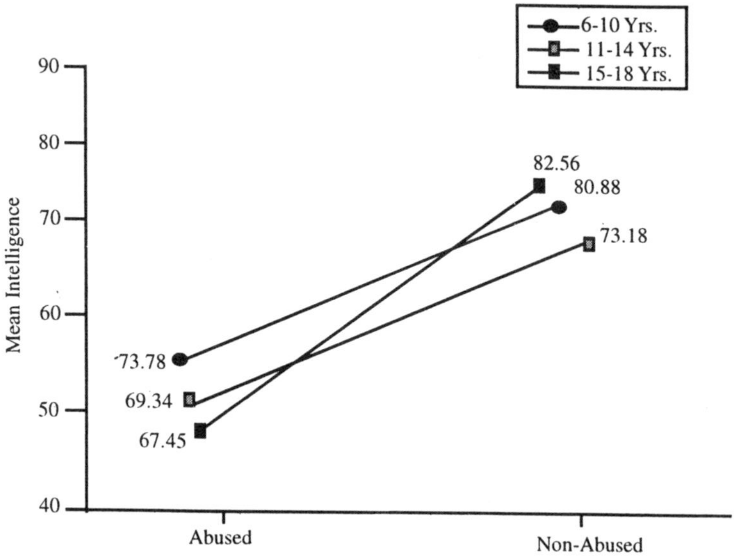

Fig. 3.7 (b) : Intelligence as a function of interaction of abuse and age

on intellectual task. Significant Child Abuse × Age interaction effect [Fig. 3.7(b)] revealed that in abused group, little variations on intellectual functioning was observed among children (M = 82.64), pre-adolescents (M = 81.81) and adolescents (M = 79.18), however, in Non-abused group, the intellectual functioning of preadolescents and adolescents were found far superior than children (M = 82.35). Findings, thus, indicated that abuse and age jointly exercised impact on intellect level of children.

(B) *Motivational Process in Relation to Child Abuse, Environmental Setting, Gender, Age and SES of Children*

The scores obtained on achievement motivation test and aspiration scale were treated statistically. Results are displayed in [Table 3.19(a), (b)] and [Figures (3.8, 3.9 & 3.10)].

Achievement Motives

(1) Achievement level Motivation of children : Results displayed in Tables 3.19 (a) & (b), suggest that abused and non-abused children

Table 3.19(a) : Shows Means, SDs and significant 'F'. values of Achievement Motivation, Aspiration, Aggression and Helping as a functions of level of abuse, environmental setting and gender.

Variables		Abused				Non-Abused				F-Values
		Rural		Urban		Rural		Urban		F_A = Child Abuse
		Boys	Girls	Boys	Girls	Boys	Girls	Boys	Girls	F_B = Environmental Setting F_C = Gender
Achievement Motivation	Mean	26.90	25.13	29.23	25.86	31.97	30.15	39.04	35.47	$F_A = 27.84$** $F_B = 7.63$**
	S.D.	12.68	13.67	14.59	14.32	13.19	11.65	9.64	10.82	$F_C = 3.54$*
Aspiration	Mean	44.86	43.38	42.00	37.36	46.07	48.82	54.92	50.16	$F_A = 35.14$** $F_{A \times B} = 12.22$**
	S.D.	11.14	13.52	15.49	14.28	11.63	8.93	8.60	12.74	$F_{B \times C} = 3.82$*
Aggression	Mean	549.94	548.77	485.82	515.21	510.20	494.73	471.29	457.69	$F_A = 26.89$**
	S.D.	87.48	71.47	76.76	93.85	70.24	67.72	44.48	52.21	$F_B = 29.47$**
Helping Behaviour	Mean	32.64	30.9	29.16	28.38	30.60	32.18	30.98	32.59	$F_A = 4.72$* $F_B = 4.60$* $F_{AXB} = 7.81$**
	S.D.	5.03	5.68	5.67	7.06	5.72	4.92	5.49	3.94	$F_{AXC} = 5.54$**

* P < .05
** P < .01

Table 3.19(b) : Shows Means, SDs and significant 'F'-values of Achievement motivation, Aspiration, Aggression and Helping as a function of level of abuse, SES and age

Variables	*Abused*									*Non-Abused*								
	Low			*Middle*			*High*			*Low*			*Middle*			*High*		
	6-10 Yrs.	*11-14 Yrs.*	*15-18 Yrs.*	*6-10 Yrs.*	*11-14 Yrs.*	*15-18 Yrs.*	*6-10 Yrs.*	*11-14 Yrs.*	*15-18 Yrs.*	*6-10 Yrs.*	*11-14 Yrs.*	*15-18 Yrs.*	*6-10 Yrs.*	*11-14 Yrs.*	*15-18 Yrs.*	*6-10 Yrs.*	*11-14 Yrs.*	*15-18 Yrs*
Achievement Motivation																		
Mean	22.94	20.03	30.24	24.39	24.48	37.44	25.86	38.9	40.43	30.9	22.29	34.15	32.21.	30.94	38.35	32.09	40.45	43.24
S.D.	12.47	13.03	13.83	12.36	13.71	8.74	17.64	6.82	3.41	15.25	13.33	9.12	12.67	13.58	7.98	8.54	6.96	5.59
F-values	FA = (Child Abuse) = 10.25**, FB (SES) = 16.13**, FC (Age) = 16.62** & FBXC = 3.85*																	
Aspiration																		
Mean	33.76	35.21	41.32	46.5	46.48	47.25	50.29	50.7	56.29	36.5	45.29	43.77	50	49.76	52.15	52.91	55	56.64
S.D.	11.31	14.32	12.81	13.87	12.35	9.83	12.84	7.97	9.86	13.03	9.87	10.92	8.8	8.94	11.9	8.82	10.3	8.3
F-values	FA= (Child Abuse) = 7.77**, FB (SES) = 37.09** & FC (Age) = 3.56*																	
Aggression																		
Mean	557.73	544.74	503.24	508.33	529.76	487.44	518.71	531.4	467	512.9	471.64	478.77	512.84	473.35	480.6	476.32	471.9	53.92
S.D.	84.32	101.85	79.85	73.4	75.47	60.45	120.03	99.9	46.95	76.98	77.19	66.81	60.44	51.72	54.31	63.64	46.73	43.34
F-values	FA (Child Abuse) = 15.23** & FC (Age) = 5.60**																	
Helping																		
Mean	29.18	29.44	30.96	31.17	31	31.06	33.43	29.4	30.43	31.6	31.14	30.69	32.79	31.94	31.25	30.91	31.55	32.44
S.D.	6.98	7.37	6.24	5.11	5	4.9	2.51	4.25	7.18	5.23	4.15	5.43	5.92	4.85	4.2	5.46	4.69	4.85
F-values	Non-Significant																	

**P<. 01, *P<. 05

differed significantly on achievement level. Abused children were found low achievers (M = 26.78) than non-abused children (M = 34.16). Achievement motive also varied significantly across Environmental Setting and Gender. Urban children were found high achievers (M =32.39) than rural children (M =28.54). Similarly, boys were found high achievers (M =31.78) than girls (M =29.15). In addition, significant main effect for SES indicates that low SES children, expressed inferior level of achievement motivation (M = 26.77) as compared to middle (M = 31.30) and high SES (M = 36.83) children. Further, adolescents showed higher achievement level (M = 37.32) than preadolescents (M = 29.52) and children (M = 28.07) respectively Table 3.19 b. Significant SES ⊗ Age interaction effect (Fig 3.8) denoted that high SES children belonging to three age groups expressed higher achievement levels, in accordance with increasing order of age of similar pattern was also observed in case of low three S.E.S. group children. However: in case of preadolescents a large variation on achievement level was found across high, middle and low SES groups. In otherwords, although high SES preadolescents

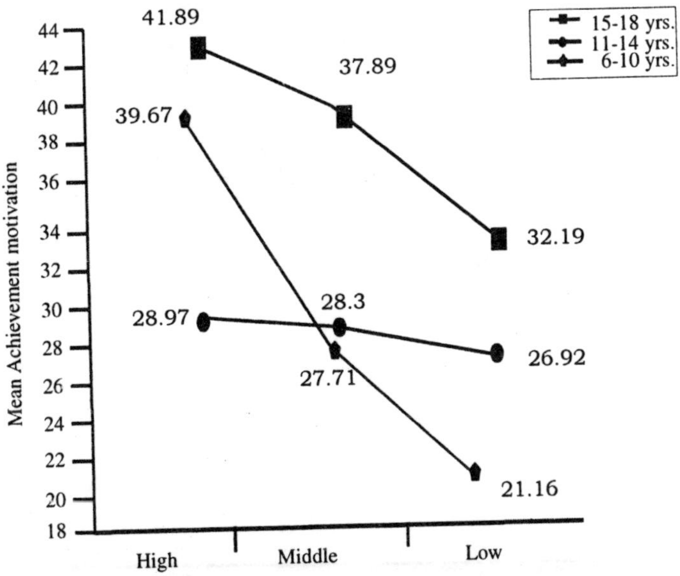

Fig. 3.8 : Achievement motivation as a function of interaction of socio-econmic status and age.

expressed higher level of achievement motive but middle and low class preadolescents showed for inferior the lowest level of achievement motive than rest of the groups.

Finding, thus, evinced that child abuse; environmental setting, gender SES and age exercised significant impact on achievement level of children.

Level of Aspiration : The level of aspiration varied significantly among abused and non-abused children. [Table-3.19(a) & (b)]. Abused children expressed low level of aspiration (M= 41.90) than non-abused children (M = 49.99). Significant Abuse × Setting interaction effect (Fig 3.9) indicated that in non-abused group, urban children expressed higher level of aspiration than rural children, contrary to this, in abused group; rural children expressed higher level of aspiration than urban

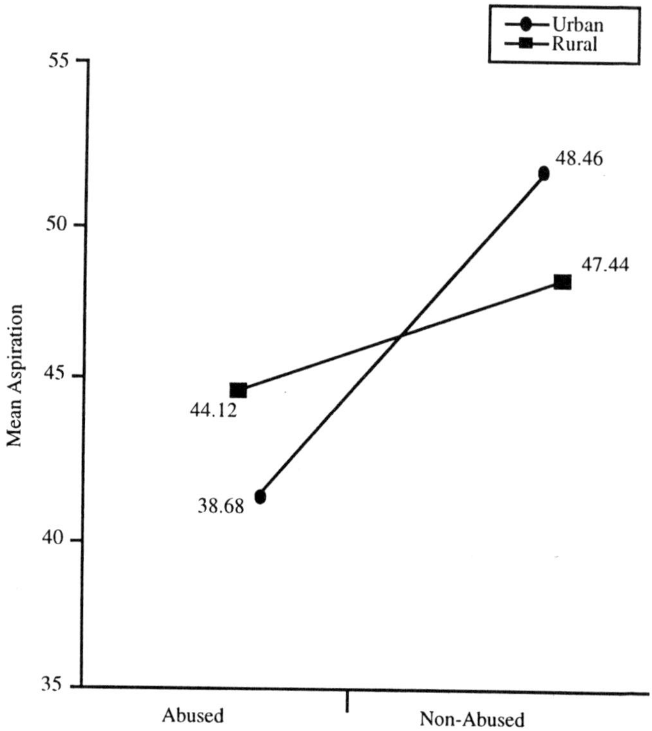

Fig. 3.9 : Aspiration as a function of interaction of the level of abuse and environmental setting.

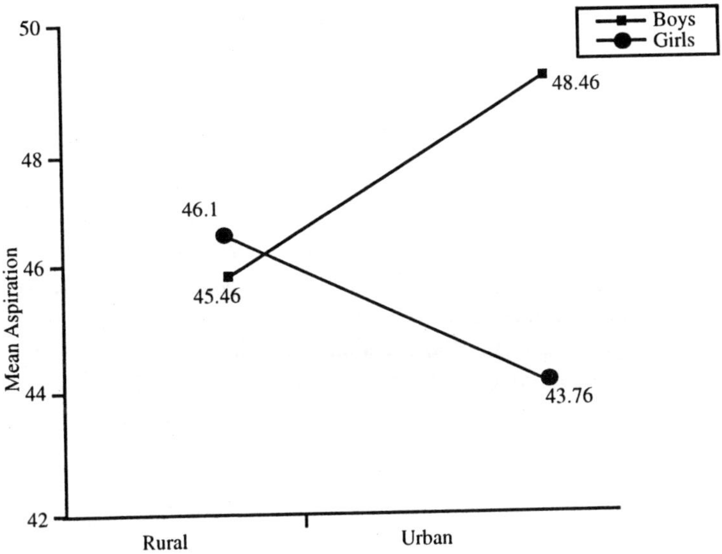

Fig. 3.10 : Aspiration as a function of interaction of environmental setting and gender.

children. Further, significant Setting × Gender interaction effect suggested that setting and gender exercised joint impact on aspiration level of children. It is evident from the Figure (3.10) that urban boys expressed higher level of aspiration (M = 48.46) than girls (M = 43.76). However, in rural sample, the differences between boys and girls were low and non significant.

(C) *Behavioural Consequences in Relation to Child Abuse, Environmental Setting, Gender, Age and SES of Children*

In order to ascertain the influence of child abuse on behavioural functioning of children, responses were obtained on helping behaviour questionnaire and aggression questionnaire. Data were scored and treated statistically. Results are displayed in Tables [3.19 (a) & (b)] and Figures (3.11 & 3.12).

(i) *Aggression in Children :* Results displayed in Table-3.19a, evinced that different groups of children expressed varying levels of aggression. Results denote that level of aggression differed significantly across abused and non-abused groups. More specifically, abused children displayed heightened level of aggression (M = 516.48) than non-abused children (M = 481.69). Further, significant main effect for Setting indicated that aggression was higher in rural children (M =529.91) than urban counterparts (M = 482.51). Significant main effect for Age denoted that younger children expressed maximum aggression (M = 514.97) than preadolescents (M = 503.80) and adolescents (M = 478.49) respectively. However, SES and gender, differences were non-significant. Findings, thus, indicate that child abuse and neglect caused heightened level of aggression in children.

(ii) *Helping Behaviour*: Helping behaviour differed significantly across abused and non-abused groups and environmental

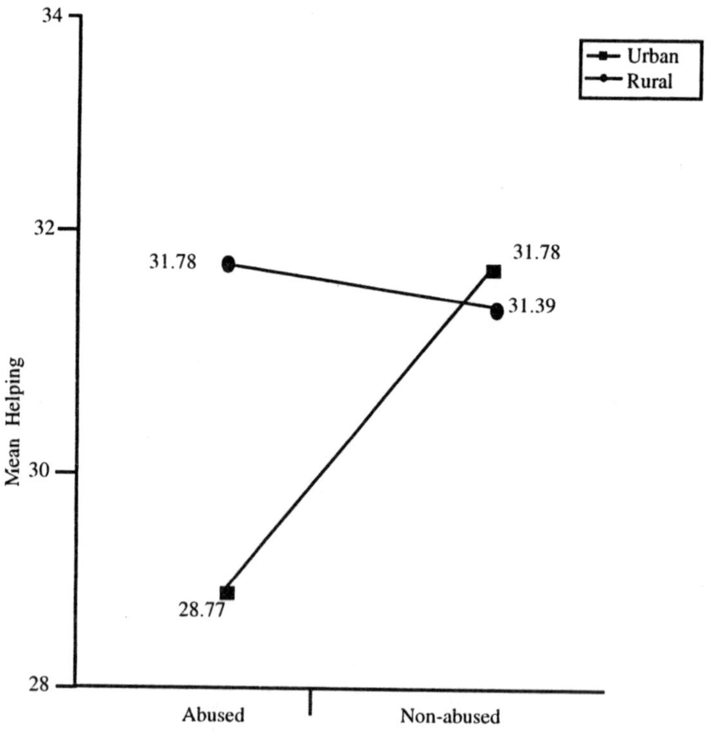

Fig. 3.11 : Helping as a function of interaction of the level of abuse and residential setting.

settings. Results, [Table 3.19(a)] indicated that non-abused children expressed more helping behaviour (M =31.59) than abused children (M = 30.27). Results further, revealed that rural children showed more helping behaviour (M = 31.58) than urban children (M = 30.28). Significant Child Abuse × Setting interaction effect (Fig 3.11), suggested that in case of abused sample, urban children showed for less helping than abused rural children but reverse pattern was found in case of non-abused sample. Urban children showed more helping behaviour than rural children. Furthermore, significant Child Abuse × Gender interaction effect (Figure 3.12) indicated

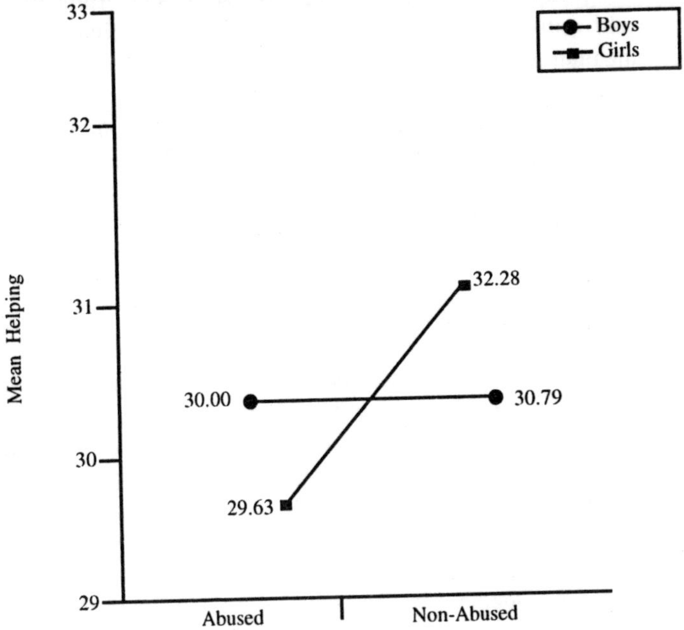

Fig. 3.12 : Helping as a function of interaction of the level of abuse and gender

that in abused group, boys showed very high level of helping behaviour than girls but in case of non-abused group, girls expressed little more helping behaviour than boys (Figure 3.12). However, SES and age differences were non-significant.

(D) *Impact of Child Abuse and Age on Personality Pattern of Children*

Raw scores obtained on CPQ and HSPQ [personality questionnaires] were transformed into Sten scores and personality profiles were prepared. Apart from this, scores were treated statistically in terms of Means, SDs and Analysis of Variance (ANOVA) for each dimension to determine the variations in personality pattern of abused and non-abused children belonging to different age groups. Results are displayed in Table 3.20 and Fig 3.13 (a) & (b). Present results denoted clear-cut variations on most of the dimensions of personality pattern of different groups of children. Although, on factor A (Warmhearted), abused and non-abused children did not differ significantly yet, significant main effect for Age indicated that pre-adolescents were found less warmhearted (M = 4.16) than younger children (M = 4.47) and adolescents (M = 4.91) respectively. On factor B, (Intelligence) abused and non-abused children varied significantly. More specifically, abused children expressed lower level of intelligence (M = 5.13) than non-abused children (M = 5.75). Furthermore, children expressed superior level of intelligence (M = 5.97) than pre-adolescents (M = 5.34) and adolescents (M = 5.01) respectively. On factor C (emotionally less stable-more stable) child abused children ware found less emotionally stable (M = 5.01) than their non-abused counter parts (M = 6.39). On Factor D (Undemonstrative *vs.* Excitable Impatient) Child Abuse varied significantly, which indicated that abused children were found highly excitable and impatient (M = 6.23) than non-abused children (M =5.80). Similarly, significant main effect for Age evinced that preadolescents were found more excitable, impatient and demanding (M = 6.67) than younger children (M = 6.42) and adolescents (M = 4.96) respectively.

On factor E (Obedient *vs.* Assertive) significant main effect for Age indicated that younger children were found more assertive (M = 7.41) than preadolescents (M = 6.47) and adolescents (M = 4.85) respectively. However, child abuse did not differ significantly.

Factor F (Tacitum *vs.* Enthusiastic) also differed significantly across three age groups. Pre-adolescents (M = 6.36) and children (M = 6.32) were found more enthusiastic and heedless than adolescents (M = 5.28).

On Factor G (Disregards *vs.* Conscientious) significant main effect for Child Abuse indicated that abused children were found more

Table 3.20 : Shows Means, SDs and significant 'F'-Values on various dimensions of personality

Variables		*Abused*			*Non-Abused*			*F - Values* *A=Child Abuse* *B= Age*
		6-10 yrs.	*11-14 Yrs.*	*15-18 Yrs.*	*6-10 yrs.*	*11-14 Yrs.*	*15-18 Yrs.*	
A	Mean	4.41	4.23	4.90	4.53	4.08	4.91	F_B= 6.69**
	S.D.	1.68	1.71	1.09	1.89	1.34	1.22	
B	Mean	5.71	4.83	4.85	6.24	5.86	5.16	F_A= 8.34**
	S.D.	2.08	1.68	1.07	2.03	2.09	1.31	F_B= 10.34**
C	Mean	4.83	4.96	5.25	5.47	5.20	8.50	F_A= 3.55
	S.D.	2.18	1.80	1.02	2.35	1.91	1.26	
D	Mean	6.52	7.14	5.02	6.31	6.20	4.90	F_A= 4.08*
	S.D.	2.23	2.11	1.20	2.34	1.88	1.43	F_B= 25.19**
E	Mean	7.22	6.87	4.90	7.59	6.08	4.81	F_B= 41.72**
	S.D.	2.05	2.22	1.50	2.37	2.49	1.50	
F	Mean	6.48	6.49	5.23	6.16	6.24	5.33	F_B= 13.34**
	S.D.	1.75	1.85	1.24	2.06	1.92	1.48	
G	Mean	4.29	4.67	5.44	5.47	5.10	5.72	F_A= 10.78**
	S.D.	2.18	1.91	1.20	1.98	1.66	1.21	F_B= 5.72**
H	Mean	4.40	4.55	5.25	4.92	5.12	5.16	F_A= 3.42*
	S.D.	1.66	1.98	1.04	1.84	1.77	1.14	F_B= 3.08*
I	Mean	4.43	5.26	5.25	5.41	4.65	5.33	N.S.
	S.D.	1.91	2.08	1.31	2.07	2.05	1.43	
J	Mean	5.55	5.78	5.00	4.88	5.06	4.97	F_A= 7.05**
	S.D.	1.75	1.81	1.58	1.77	1.66	1.02	

Contd..

Table 3.20 *(Condt..)*

N	Mean	6.64	4.83	0.4	6.06	3.33	0.22	$F_A = 7.05^{**}$
	S.D.	2.03	3.36	0.32	2.27	3.72	0.15	$F_B = 151.43^{**}$
O	Mean	6.19	5.93	5.04	5.86	5.37	4.98	$F_B = 28.01^{**}$
	S.D.	2.08	1.93	1.07	1.78	2.05	1.36	
Q2	Mean	0.07	1.28	5.10	0.12	2.65	4.90	$F_A = 4.11^{**}$
	S.D.	0.52	0.48	1.69	0.09	2.11	1.51	$F_B = 199.18^{**}$
Q3	Mean	3.93	4.48	5.04	4.49	4.78	5.09	$F_B = 5.59^{**}$
	S.D.	1.77	2.04	1.35	1.71	2.29	1.42	
Q4	Mean	6.88	6.68	5.54	6.31	6.02	5.14	$F_A = 7.39^{**}$
	S.D.	2.35	1.94	0.85	2.14	2.01	0.93	$F_B = 14.36^{**}$

* $P < .05$
** $P < .01$

expedient (disregard rules) and showed weak super ego strength (M = 4.72) than non-abused children (M = 5.07). Furthermore, this tendency varied significantly across three age groups. Pre-adolescents (M = 4.83) and children (M = 4.66) were found more expedient than adolescents (M = 5.58).

Different Groups of children also differed significantly on Factor H (Shy *vs.* Adventurous). Abused children were found more shy, timid and diffident (M = 4.73) than non-abused children (M = 5.43). Further significant main effect for Age indicated that children were found more shy and diffident (M = 4.66) than preadolescents (M = 4.88) and adolescents (M = 5.20) subsequently.

Though, different groups of children did not differ significantly on Factor I, (Tough-minded *vs.* Tender-Minded), yet on Factor J (Zestful *vs.* Circumspect-individualism), significant main effect for Child Abuse suggested that abused children were found Zestful (M = 4.97) than non-abused children (M = 5.44).

Child Abuse and Age exercised significant impact on Factor N, (Forthright *vs.* Shrewd). Abused children were found more artless, sentimental and natural (M = 3.95) than non-abused children (M = 3.21). This tendency differed significantly across three age groups. Children were found more practical and artful (M = 6.35) than pre-adolescents (M = 4.08) and adolescents (M = 0.31).

On Factor 'O' (Self assured *vs.* Apprehensive), Child Abuse did not differ significantly but significant main effect for Age suggested that adolescents were found less apprehensive (guilt prone) (M = 5.01) than preadolescents (M = 5.65) and children (M = 6. 03).

On Factor Q_2 (Group Dependent *vs.* Self-Sufficient), different group of children varied significantly. More specifically, abused children were found more group dependent and sound followers (M = 2.15) than non-abused children (M = 2.55). Significant main effect for Age indicated that group dependency was minimum in adolescents (M = 5.00) than pre-adolescents (M =1.96) and children (M = .09). It appears that level of abuse damaged the decision-making and self-sufficiency in children.

On Factor Q_3 (Uncontrolled *vs.* Controlled), different age group of children varied significantly. Adolescents were found more controlled (M = 5.06) whereas children (M = 4.21) and pre-adolescents (M =4.63) were found uncontrolled and undisciplined.

Similarly on Factor Q_4 (Relaxed *vs.* Tense) children differed significantly across level of abuse and age. Abused children were found more tense and frustrated (M = 6.37) than non-abused children

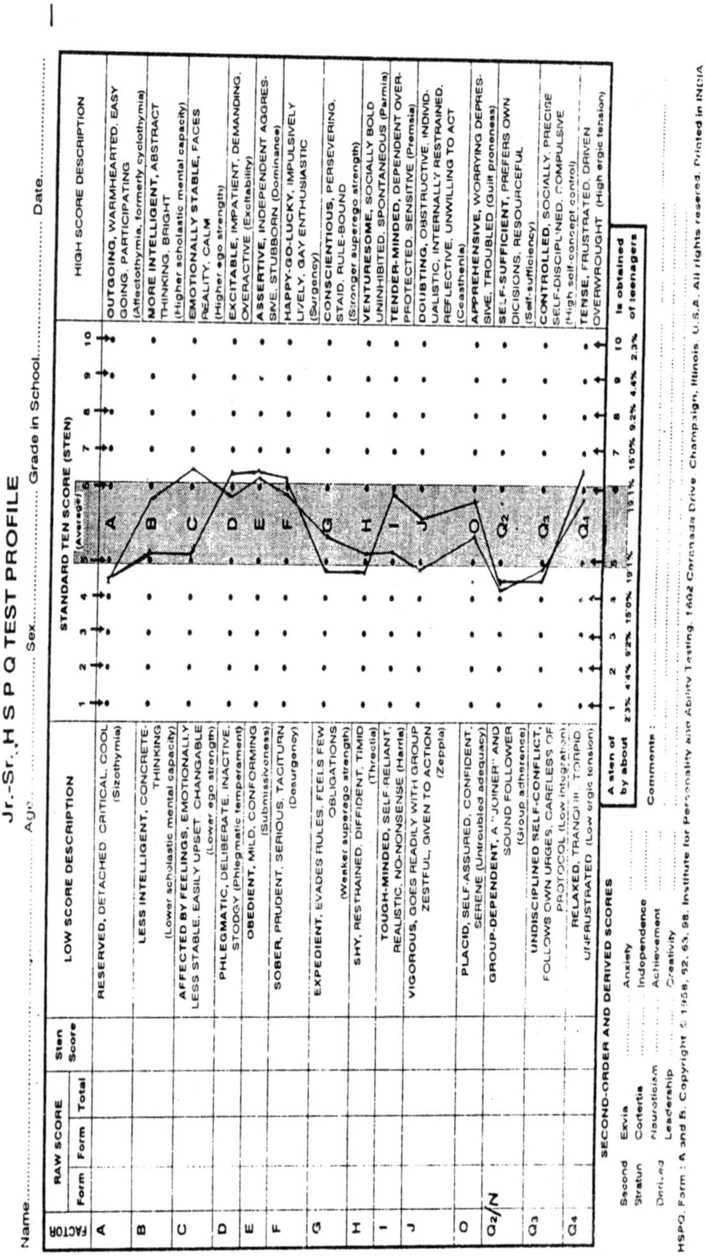

Fig. 3.13 (a) : Personality Profile of abused and non-abused children

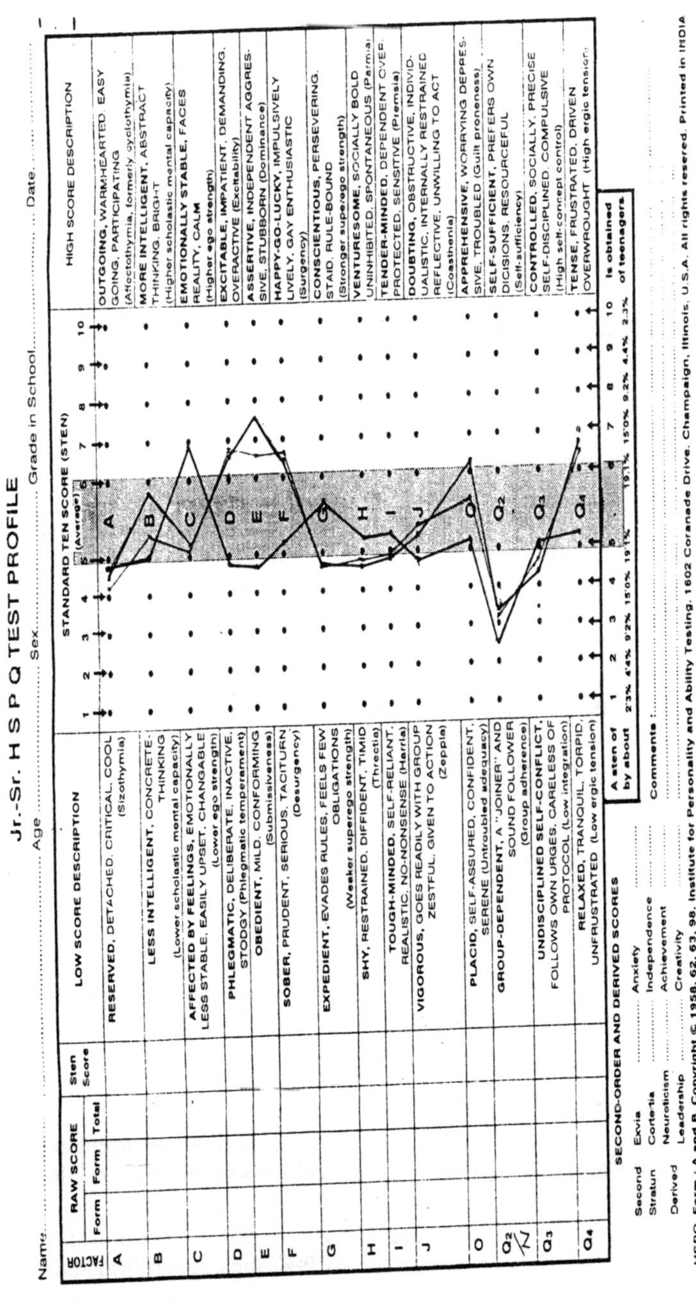

Fig. 3.13 (b) : Personality profile of different age group of children

(M = 5.82). Further, children were found more tense (M = 6.60) than pre-adolescents (M = 6.35) and adolescents (M = 5.34).

Findings, thus, suggested that abused children as compared to non-abused counterparts expressed low level of intelligence, high impatience, disregarding rules, weak super ego and were also found more impractical shy and timid, but showed low level of vigor, poor decision making ability high andfrustration. Result, thus, indicated that abusive treatments by parents or any significant others damage proper personality development of children.

Correlational Analysis

In order to obtain further insight into the linkages among variables, correlational analysis was done for child sample. Correlations were computed to ascertain :

(1) The relationship of child abuse and neglect with various forms of cognitive functioning, viz. perceptual identification, verbal learning, memory span, creativity and intellectual functioning;
(2) The relationships of child abuse and neglect with achievement motivation and level of aspiration;
(3) The linkages of child abuse and neglect with aggression and helping behaviour, and
(4) The relationships of child abuse and neglect with various dimensions of personality.

(1) *Relationships of Child Abuse and Neglect with Cognitive Functioning*

Correlational analyses were done to establish the associations of abuse and neglect with perceptual identification, verbal learning and memory, creative thinking and intellectual functioning. Results are displayed in Table 3.21 and interpreted below.

(i) *Perceptual Identification :* Table 3.21 denotes an inverse relationship between abuse and perceptual identification. Similarly, negative relationship between child neglect and perceptual identification was found.
(ii) *Verbal Learning and Memory Span* : Correlation results (Table 3.21) denote, negative relationship of child abuse with

verbal organization (clustering). Similarly, child neglect was also found inversely related with verbal organization. Table 3.21 further indicates that child abuse and neglect were found inversely related with memory span.

(iii) *Creativity:* Correlations displayed in Table 3.21, further denoted that child abuse was found inversely correlated with verbal fluency, verbal flexibility, verbal originality, non-verbal related elaboration for picture, elaboration for title, nonverbal related originality for picture, originality for title and non-verbal flexibility.

Results, further, indicated that neglect was also found to be negatively related with, verbal fluency, verbal flexibility, verbal originality, non-verbal related elaboration for picture, elaboration for title, non-verbal related originality for picture, originality for title and non-verbal flexibility.

(iv) *Intellectual Functioning:* Results Table 3.21 indicated highly negative relationship between child abuse and intelligence as well as between child neglect and intellect of children.

Findings, thus, suggest that child abuse and neglect have inverse associations with various forms of cognitive functioning.

Table 3.21 : Shows relationships of Child Abuse and Neglect with Perceptual Identification, Verbal Organization (Clustering), Memory Span, Creative Thinking, and Intellectual Functioning

		Variables Name	*Abuse (r)*	*Neglect (r)*
		Perceptual Identification (EFT)	-.27**	-.24**
		Verbal Organization (Clustering)	-.26**	-.26**
		Memory Span	-.38**	-.38**
Dimensions of Creative Thinking	Creative (Verbal) Thinking	Fluency (V)	-.25**	-.27**
		Flexibility (V)	-.27**	-.30**
		Originality (V)	-.26**	-.25**
	Creative (Non-Verbal) Thinking	Elaboration (N) (Picture)	-.31**	-.32**
		Elaboration (V) (Title)	-.29**	-.31**
		Originality (N) (Picture)	-.27**	-.30**
		Originality (V) (Title)	-.25**	-.22**
		Flexibility (Non-Verbal)	-.25**	-.27**
		Intelligence	-.28**	-.23**

N= 335, * P< .05, ** P < .01

(2) Linkages of Child Abuse and Neglect with Motivational Processes

Correlations were also computed to ascertain the relationships of abuse/neglect with achievement motivation and level of aspiration.

(i) *Achievement Motivation*: Results Table 3.22 indicated an inverse relationship between child abuse and achievement motivation of children. Further, highly negative relationship between child neglect and achievement motive, was identified.

(ii) *Level of Aspiration:* Results displayed in Table 3.22, further, reveals an inverse relationship between child abuse and level of aspiration. Similarly, neglect was found to be negatively related with level of aspiration. It is apparent from this result, that child abuse and neglect disrupt the motivation level of children.

Table 3.22: Shows the relationships of Child Abuse and Neglect with Achievement Motivation, Aspiration, Aggression, Helping and Personality Pattern of children

Variables Name	*Abuse (r)*	*Neglect (r)*
Achievement Motivation	-.33**	-.34**
Aspiration	-.42**	-.46**
Aggression	.24**	.215**
Helping	-0.07	-0.07
Personality factors		
A	-0.03	-0.04
B	-.14**	-.16**
C	-.195**	-.17**
D	.17**	.165**
E	0.06	0.09
F	0.08	0.08
G	-.215**	-.19**
H	-0.1	-.13*
I	-0.03	-0.03
J	.185**	.22**
N	.175**	.13*
O	.11*	.11*
Q_2	-.12*	0.1
Q_3	-.11*	-0.06
Q_4	.20**	.16**

* P< .05
** P< .01

(3) *Relationship of Child Abuse and Neglect with Behavioural Functioning*

To determine the association of abuse and neglect with behavioural processes of children, correlations were computed and results are displayed in Table 3.22.

(i) *Aggression :* Correlations displayed in Table 3.22 evinced highly positive relationship between child abuse and aggression. Similarly, child neglect was also found positively correlated with aggression.

(ii) *Helping Behaviour :* Correlation results (Table 3.22) further, indicate that child abuse was found negatively correlated with helping behaviour. Similarly, child neglect and helping behaviour were found negatively correlated with each other, although correlation values were found low and negative.

(4) *Linkages of Child Abuse/Neglect with Personality Pattern of Children*

Correlations (Table 3.22) denoted different patterns of relationships of abuse and neglect with various factors of personality. Although, relation with Factor A was non-significant but child abuse was found to be inversely correlated with Factor B (Intelligence), Factor C (emotional stability high-superego), Factor G (conscientious), Factor Q_2 (socially-group dependent *vs.* self-sufficient) and Factor Q_3 (uncontrolled *vs.* controlled).

Contrary to this, abuse was found positively correlated with Factor D (excitability), Factor J (circumspect, individualism), Factor N (shrewdness), Factor O (apprehensive or guilt prone, self sufficient) and Factor Q_4 (tense).

Table 3.22, further denotes that child neglect was found negatively associated with Factor B (intelligence), Factor C (emotional stability, high superego), Factor G (conscientious), Factor H (adventurous), whereas, it was found positively correlated with Factor D (excitability), Factor J (circumspect), Factor N (shrewdness), Factor O (apprehensive or guilt prone, self-sufficient) and Factor Q_4 (Tense).

Present results thus suggested that positive traits of personality were found negatively linked with child abuse and neglect, however, negative traits were found positively associated with child abuse and neglect. Findings, thus, evinced the ill consequences of abuse and neglect on personality development of children.

Multiple Regression Analysis

Prediction of Consequence Variables from Child Abuse and Neglect

The correlations, reported in earlier section of the result showed that child abuse and neglect were found negatively correlated with cognitive, motivational and behavioural functioning of children. The close linkages between child abuse/neglect and Personality factors were also identified. Hence, it was considered relevant to go for stepwise multiple regression (SMRA) analysis to determine the relative contributions of child maltreatment (Predictors) to criterion variables. Regression results are displayed in Table No. 3.23, 3.24, 3.25 & 3.26) and interpreted in the following section.

1. Prediction of Cognitive Processes from Child Abuse and Neglect

Based on the perusal of correlation results, it was viewed to go for stepwise multiple regression analysis (SMRA) to determine the relative contributions of child abuse and neglect to perceptual identification, verbal clustering, memory span, creativity and intelligence. The findings of this analysis are in order :

(i) *Perceptual Identification:* As Table 3.23 denotes,Perceptual Identification was predicted by Child Abuse as a whole and Child Educational Neglect. Child Abuse contributed maximum negatively (β = -.27, R^2=.07, df, 1,333, P<.00) followed by Child Educational Neglect (β = -.16, R^2=.087, df, 2,332, P<.05).

(ii) *Verbal clustering :* Result revealed that clustering was explained by two factors. Child Physical Abuse contributed maximum negatively (β =-.32, R^2=.10, df, 1, 333, P<.00), followed by Abuse as a whole (β = -.25, R^2=.12, df, 2,332, P<.00).

(iii) *Memory Span :* The Memory Span of children was predicted by four forms of child maltreatments. Child Educational Neglect contributed maximum negatively (β=-.39, R^2 =.152, df, 1,333, P<.00), followed by Child Physical Abuse

(iii) *Memory Span* : The Memory Span of children was predicted by four forms of child maltreatments. Child Educational Neglect contributed maximum negatively (β=-.39, R^2 =.152, df, 1,333, P<.00), followed by Child Physical Abuse (β = -.22, R^2=.18, df, 2,332, P<.00), Psychological Abuse (β = -.14, R^2 =.19, df, 3,331, P< .05) and Abuse as a whole (β = -.77, R^2=.21, df, 4,330, P<.01) respectively.

(iv) *Creative Potentials*: The Regression result Table (3.24) further, evinced that various dimensions of creativity (Verbal and Non-verbal) were negatively predicted by child abuse and neglect. Results revealed that :

- The Verbal Fluency was predicted negatively by Child Neglect as a whole, (β = -.27, R^2=.07, df, 1,333, P<.00),
- The Verbal Flexibility was also explained negatively by Child Neglect as a whole, (β = -.296, R^2=.09, df, 1,333,

Table 3.23 : Step-wise Multiple Regression Analysis of Cognitive Functioning i.e., Perceptual Identification, Verbal Organization (Clustering), Memory Span and Intelligence on to the various forms of Child Abuse and Neglect.

Predictors	*Perceptual Identification*				
	R	*R²*	*R² change*	*Beta*	*F*
Abuse as a Whole	.27	.07	.07	-.27	26.47**
Educational Neglect	.30	.09	.01	-.16	15.86*
	Verbal Organization (Clustering)				
Physical Abuse	.32	.10	.10	-.32	37.25**
Abuse as a Whole	.35	.12	.02	-.25	22.46**
	Memory Span				
Educational Neglect	.39	.15	.15	-.39	59.77**
Physical Abuse	.43	.18	.03	-.22	36.81**
Psychological Abuse	.44	.19	.01	-.14	26.18**
Abuse as a Whole	.46	.21	.02	-.77	21.92**
	Intelligence				
Neglect as a Whole	.38	.15	.15	-.38	56.57**
Psychological Abuse	.41	.17	.02	-.21	32.80**

** P<.01
*P<.05

Table 3.24 : Step-wise Multiple Regression Analysis of Creative Thinking on to the various forms of Child Abuse and Neglect.

Creative Thinking *(Verbal)*

Predictors	*Fluency*				
	R	R^2	R^2 *change*	*Beta*	*F*
Neglect as a Whole	.27	.07	.07	-.27	25.27**
		Flexibility			
Neglect as a Whole	.30	.09	.09	-.30	39.95**
		Originality			
Psychological Abuse	.28	.08	.08	-.28	27.89**

Creative Thinking (*Non-Verbal)*

Predictors	*Elaboration (n)*				
	R	R^2	R^2 *change*	*Beta*	*F*
Neglect as a Whole	.32	.10	.10	-.32	37.48**
Psychological Abuse	.34	.12	.01	-.17	21.56**
		Elaboration (v)			
Physical Abuse	.37	.14	.14	-.37	52.02**
Physical Neglect	.40	.16	.03	-.20	31.66**
		Originality (n)			
Physical Abuse	.31	.10	.10	-.31	35.75**
Neglect as a Whole	.33	.11	.01	-.17	20.77**
Sexual Abuse	.36	.13	.02	-.13	16.06*
		Originality (v)			
Psysical Abuse	.26	.07	.07	-.26	24.27**
		Flexibility (Non verbal)			
Physical Abuse	.39	.15	.15	-.39	58.76**
Sexual Abuse	.42	.18	.03	.17	36.09**

** P<.01
*P<.05

- Further, Elaboration-N (for picture) was maximum explained by Neglect as a whole negatively (β = -.32,

R^2=.10, df, 1,333, P<.00), followed by Psychological Abuse (β = -.17, R^2=.115, df, 2,332, P<.05),

- The Elaboration V (for title) was predicted by two factors. Child Physical Abuse contributed maximum negatively (β =-.37, R^2=.135, df, 1,333, P<.00), followed by Physical Neglect (β = -.196, R^2=.16, df, 2,332, P<.00),
- Further, the Originality N (for picture) was explained by three factors. Child Physical Abuse (CPA) predicted maximum negatively (β =-.31, R^2=.097, df, 1,333, P<.00), followed by Neglect as a whole (β =-.166, R^2=.11, df, 2,332, P<.05) and Sexual Abuse (β =-.13, R^2=.13, df, 3,331, P<.05) respectively,
- The Originality V (for title) was negatively explained by Child Psychological Abuse (β = -.26, R^2=.07, df, 1, 333, P<.00) and ,
- Non-verbal Flexibility was contributed by physical abuse and sexual abuse. Child Physical Abuse (CPA) explained maximum negatively (β = -.387, R^2=.15, df, 1,333, P<.00) followed by Sexual Abuse (β=-.17, R^2=.18, df, 2, 332, P<.001).

(v) *The Intellect* : Regression result further indicated that the level of Intelligence was contributed by Child Neglect as a whole and Psychological Abuse (C Psy A). The Child Neglect predicted maximum negatively (β = -.38, R^2=.145, df, 1,333, P<.00) followed by Child Psychological Abuse (β = -.21, R^2=.165, df, 2, 332, P<.01).

Regression results, thus, indicated that various forms of Child Abuse and neglect were identified as strong predicators of Perceptual, Learning, Memory, Creativity and Intellectual processes. It appears that maltreating experiences interrupt the proper development of cognitive talents in children.

2. *Prediction of Motivational Processes from Child Abuse and Neglect*

The Stepwise Multiple Regression Analysis was also done to explore the relative contributions of Child Abuse and Neglect to levels of Achievement Motive and Aspirations in children. Results displayed in Table 3.25 evinced that :

- Achievement Motivation was predicted by two factors.

The Child Physical Abuse contributed maximum negatively (β = -.357, R^2=.13, df, 1,333, P<.00) followed by Neglect as a whole (β = -.17, R^2=.14, df, 2,332, P<.05).

- Similarly, Aspiration in children was explained by three factors. Child Neglect as a whole contributed maximum negatively (β =-.46, R^2=.21, df, 1,333, P<.00), followed by Physical Neglect (β =-.24, R^2=.237, df, 2,332, P<.002) and Physical Abuse (β =-.146, R^2=.25, df, 3,331, P<.05) respectively.

It is apparent from the result that abusive treatments damage the motivation level of children.

3. *Prediction of Behavioral Functioning (Aggression and Helping Behavior)*

In order to determine the contribution of child abuse and neglect to Aggression and Helping Behaviour (criterion variables), Stepwise Multiple Regression Analysis was calculated and results of the analysis (Table 3.25) are in order:

Table 3.25 : Step-wise Multiple Regression Analysis of Achievement Motivation, Aspiration, Aggression and Helping Behaviour on to the various forms of Child Abuse and Neglect.

Predictors	*Achievement Motivation*				
	R	*R^2*	*R^2 change*	*Beta*	*F*
Physical Abuse	.36	.13	.13	-.36	48.61**
Neglect as a Whole	.38	.14	.02	-.17	27.56**
		Aspiration			
Neglect as a Whole	.46	.21	.21	-.46	90.66**
Physical Neglect	.49	.24	.02	-.24	51.52**
Physical Abuse	.50	.25	.01	-.15	36.32*
		Aggression			
Physical Abuse	.24	.06	.06	-.24	20.98**
		Helping Behaviour			
Physical Neglect	.15	.02	.02	-.15	7.23**

** P<.01
*P<.05

- The Aggression in children was predicted by Child Physical Abuse (CPA), positively (β = .24, R^2=.06, df, 1,333, P<.00), whereas; Helping behaviour was explained by Child Physical Neglect, negatively (β = -.146, R^2=.02, df, 1, 333, P<.00). Results thus, evinced that child abuse heightened the aggression level in children and diminished their helping tendencies.

4. *Predictions of Personality Factors from Child Abuse and Neglect*

Correlation results, mentioned in earlier section of the result also evinced that child abuse and neglect were found closely related to different dimensions of personality. Based on the perusal of correlation results, it was considered important to go for Stepwise Multiple Regression Analysis (SMRA) to examine the relative contributions of Child Abuse and Neglect on different dimensions of Personality. The regression results :

Results revealed that although, Factor A (Reserved *vs.* Outgoing) was not explained by any factor, however Factor B, (Less Intelligence *vs.* More Intelligence) was predicted by Child Emotional Neglect (β = -.16, R^2 =.025, df, 1, 333, P<.01). The Factor C (Less Emotional Stability *vs.* More Emotional Stability) was explained by Child Physical Abuse (CPA), negatively (β = -.233, R^2=.054, df, 1, 333, P<.00). Contrary to this, Factor D (Phlegmatic *vs.* Excitable) was positively predicted by Child Physical Abuse (CPA) (β =-.257, R^2=.066, df, 1, 333, P<.00) followed by Child Sexual Abuse, which contributed negatively. (β = -.177, R^2=.096, df, 2,332, P<.01). Results further indicated that Factor E (Obedient *vs.* Assertive) was explained by three factors. Child Physical Abuse (CPA) contributed maximum positively. (β = .18, R^2=.03, df, 1,333, P<.001) followed by Child Sexual abuse which contributed negatively (β=-.21, R^2=.08, df, 3,331, P<.05). The Factor F (Sober *vs.* Happy-go-lucky) was also maximum positively contributed by Child physical abuse (β = .135, R^2=.018, df, 1,333, P<.01) and Child Sexual Abuse, which contributed negatively. (β= -.16, R^2=.04, df, 2,332, P<.01). Factor G (Expedient *vs.* conscientious) was contributed by two factors. Child physical Abuse (CPA) predicted maximum negatively (β= -.325, R^2=.105, df, 1,333, P<.00) followed by Abuse as a whole (β= -.20, R^2=.12, df, 2,332, P<.05). Apart from this, Factor H (Shy *vs.* Venturesome) was negatively

Table 3.26 : Step-wise Multiple Regression Analysis of Personality Factors on to the various forms of Child Abuse and Neglect.

Predictors	*R*	*R^2*	*R^2 change*	*Beta*	*F*
		Factor-B			
Emotional Neglect	.16	.03	.03	-.16	8.61*
		Factor-C			
Physical Abuse	.24	.05	.05	-.24	19.19*
		Factor-D			
Physical Abuse	.26	.07	.07	.26	23.59**
Sexual Abuse	.31	.10	.03	-.18	17.65**
		Factor-E			
Physical Abuse	.18	.04	.04	.18	11.48**
Sexual Abuse	.27	.07	.04	-.21	13.01**
Abuse as a Whole	.29	.08	.02	-.21	10.06*
		Factor-F			
Physical Abuse	.14	.02	.02	.14	6.23*
Sexual Abuse	.21	.04	.03	-.16	7.43**
		Factor-G			
Physical Abuse	.33	.11	.11	-.33	39.26**
Abuse as a Whole	.34	.12	.02	.21	22.04*
		Factor-H			
Educational Neglect	.15	.03	.03	-.15	7.35**
		Factor-J			
Physical Abuse	.25	.06	.06	.25	21.58**
		Factor-N			
Physical Abuse	.28	.08	.08	.28	28.77**
Sexual Abuse	.37	.14	.06	-.25	26.43**
		Factor-O			
Physical Abuse	.17	.03	.03	.17	10.51**
Sexual Abuse	.22	.05	.02	-.13	7.85*
		Factor-O^2			
Physical Abuse	.26	.07	.07	-.26	24.72**
Sexual Abuse	.40	.16	.09	.31	31.61**
		Factor-O^3			
Physical Abuse	.15	.02	.02	-.45	7.15**
		Factor-O^4			
Physical Abuse	.25	.06	.06	.25	21.61**

** P<.01

*P<.05

explained by Child Educational Neglect (β= -.147, R^2=.02, df, 1,333, P<.01). However, Factor I (Vigorous *vs.* Doubting) was contributed by Child Physical Abuse (CPA) (β = .25, R^2=.06, df, 1, 333, P<.00), and Child Sexual Abuse, which contributed negatively (β = -.245, R^2=.14, df, 2,332, P<.00). The result, further, evinced that Factor O, (Placid *vs.* Apprehensive) was predicted by two forms of child maltreatment. Child Physical Abuse (CPA) explained maximum positively (β= .175, R^2=.03, df, 1,333, P<.001) followed by Child Sexual Abuse which contributed negatively (β= -.123, R^2= .045, df, 2,332, P<.05). Similarly, Factor Q_2 (Group Dependent *vs.* Self Sufficient) was contributed by two factors. The Child Physical Abuse explained maximum negatively (β= -.263, R^2=.07, df, 1, 333, P<.00) followed by Child Sexual Abuse which contributed positively (β= -.31, R^2=.16, df, 2,332, P<.00). Factor Q_3 (Undisciplined *vs.* Controlled) was also explained by Child Physical Abuse negatively (β = -.15, R^2=.134, df, 2,332, P<.01), Further, Q_4 (Relaxed *vs.* Tense) was predicted by child Physical Abuse positively. (β= -.247, R^2=.02, df, 2,333, P<.00). Present results, thus, evinced that abuse and neglect exercised negative role in the proper development of personality.

In brief, results recorded in various sections of this chapter evince that child abuse is caused by multiple factors. In this process, demographic characteristics of parents, children and family play prominent roles. In addition, psychological dispositions also have defect links with child abuse and neglect. Furthermore, child abuse in turn exerts adverse impact on Cognitive, multinational, behaviour functionings and personality pattern of children.

4

INTERPRETATION AND DISCUSSION

The findings of present study are interpreted and discussed under five major sections. At the out set, major findings of the study have been interpreted in accordance with the statistical treatments, presented in result section. In the subsequent four sections, the role of predicting variables in the origin of child abuse and neglect and their impacts on various psychological processes and personality pattern of children discussed in the light of empirical and theoretical evidences.

The first section, deals with the role of environmental and demographic factors in the origin of child abuse and neglect. In the second section, the social status (SES and caste) and occupational characteristics of abusive and non-abusive families are interpreted. Third section of the chapter deals with some psychological dispositions, i.e., early abusive experiences and attitudes towards parents, stress, anxiety and coping in abusive and non-abusive parents/adults and the relationships of these factors with child abuse and neglect, and Fourth section of the chapter focuses on the impacts of child abuse and neglect on cognitive functioning, motivational and behavioural processes and personality pattern of children and lastly a model has been developed and depicted linking antecedents and consequences of child abuse. Further, some of the issues for future researches and implications of the study have been discussed in the light of present piece of research.

MAJOR FINDINGS OF THE STUDY

Findings of this research suggested that the magnitude of child maltreatment (abuse and neglect) varied in accordance with variations in environmental, socio-demographic and parental characteristics. Further, child abuse exercised negative impact on cognitive, motivational and behavioural processes and personality pattern of children. Findings are interpreted in brief, according to objectives of the study.

Findings of Objective - I

1. ANOVA results reported that rural children were found more abused and neglected than their urban counterparts on some forms of abuse and neglect. The demographic characteristics of children, i.e., age and gender were found potential contributors to child abuse and neglect. Maximum physical abuse was found in pre-adolescents, whereas a sizeable number of adolescents were found the victims of sexual abuse. Various forms of neglects, viz. physical, emotional and educational, were found significantly greater in girls than their boys counterparts. Illiterate or low educated children were found more victims of abuses and neglects than highly educated children. The extent of abuse also differed across birth order of the children. Later born were found more abused and neglected than early or first-born children.

Certain family characteristics were also found important predictors of abuse and neglect. The levels of abuse and neglect varied as a function of family structure, i.e. type of family, family size (number of family members and number of children) and family environment. Results show that children belonging to nuclear families were found more abused and neglected than those of joint families. More specifically, it was found that :

- The extent of child abuse and neglect were significantly higher in larger families as compared with small size families,
- The different forms of abuse and neglect were found greater in residents (children) of small size houses than those of large size houses, and
- The family environment was also found an important contributor to abuse and neglect. Abused children reported poor levels of cohesion, intellectual-cultural, orientation, expressiveness, independence, achievement orientation, active recreational orientation, moral religious emphasis and organization but higher, control and conflict in their family environment.

Results further indicated that certain demographic characteristics of parents, i.e. parental age, gender, education, health status also exercised impact on child abuse and neglect. It was found that:

- Younger parents were found more abusive than older parents,
- Mothers were found more neglectful than fathers,
- Illiterate/low educated parents were found more abusive and neglectful than highly educated parents,
- Handicapped or disabled parents were found more abusive and neglectful than weak and normal health parents subsequently.

2. Correlation results indicated that child abuse and neglect were found to be negatively related with child's age, education, birth order and health status but positively related with gender, which suggested that with increasing child's age, education, and health status, the levels of abuse and neglect decreased. However, positive relationships with gender and child birth order suggested that girls as well as later born were found more maltreated than boys and first born or only children.

Certain family characteristics were found negatively related with child abuse and neglect. More specifically, child abuse and neglect were found negatively related with type of family, size of family and size of house. Hence, child abuse and neglect were found greater in nuclear family, larger family, and small size houses than the comparison groups.

Family environment was also found exclusively related to child abuse and neglect. Cohesion, expressiveness, independence, achievement orientation, intellectual cultural orientation, active recreational orientation, moral religious emphasis and organization were found inversely related to child abuse and neglect, however, negative domains, i.e., conflict and control were positively related with child maltreatment.

Apart from this, child abuse and neglect were found to be negatively related with parental age, education and health status, but positively related with gender. Findings, thus, suggested that with increasing age, education and health status, parents were found less abusive and neglectful. However, mothers were found more neglectful than fathers.

3. The findings of Stepwise Multiple Regression Analysis (SMRA) evinced that demographic characteristics of children, parents and family contributed significantly to the incidences of various forms of abuse and neglect. Present results evinced that :

(i) Child Physical Abuse (CPA) was positively predicted by small size of house, low education, lack of independence in family environment, poor health status of child, younger age of parents, rural environment and nuclear family structure respectively;

(ii) Child'Psychological Abuse (CPA) was positively explained by small size of house, poor health status of child, nuclear family, low cohesion in the family and poor health of parents subsequently;

(iii) The Child Sexual Abuse (CSA) was favorably contributed by adolescence age and insufficient living space in the house (small size of house), and

(iv) Abuse as a whole was favorably predicted by small size of house, poor health, and low cohesion in family, nuclear family structure, illiteracy or low education of child, younger age, poor organization and low independence in the family environment respectively.

It is also evident from regression results that child neglect was also contributed by numerous demographic factors. It was found that:

(i) Higher education of parents and children predicted low incidences of Child Physical Neglect, however, younger age and gender (girls) explained high incidences of such neglects,

(ii) Child Emotional Neglect was positively explained by small size of house, child's poor health status, low education of parents, high conflict in family, gender (mother), nuclear family and poor health condition of parents respectively,

(iii) Child Educational Neglect was explained by various factors, i.e. low education, improper living space, poor health status, nuclear family structure, high independence in family environment, later birth order and elder age of children respectively, which predicted higher rates of child educational neglect, and

(iv) Neglect as a whole was favorably predicted by illiteracy/low education level of parents, nuclear family structure and age of children.

Findings of Objective - II

1. ANOVA results obtained on second objective indicated that social status (socio-economic status, caste and community) and occupational

characteristics exercised impact on child abuse and neglect. More specifically results denoted that :

- Different forms of abuses and neglects were higher in low caste (SC/ST) than backward and upper caste families subsequently.
- Child maltreatment was found more in Muslim community than Hindu community.
- The victims of child abuse and neglect were found very high in low socio-economic status families than middle and high socio- economic status groups respectively.
- Although, the cases of child abuse and neglect were identified in the entire occupational groups in some or other forms but maximum abuses and neglects were identified in labors than cultivators, business and service class respectively.

2. Correlation results also denoted that child maltreatment (abuse and neglect) was found to be negatively related with caste and community, socio-economic status and occupation. Findings, thus, denoted that with increasing the level of caste, socio-economic status and occupation, the extent of abuse and neglect decreased.

3. The SMRA results facilitated the prediction of child abuse and neglect. Findings suggested that Child Physical Abuse (CPA), Psychological Abuse, Child Sexual Abuse (CSA) and Abuse as a whole were favorably explained by low status (socio-economic status, caste and community) and lower occupations of parents/families respectively.

Similarly, Child physical neglect, Emotional neglect, Educational neglect and Neglect as a whole were also positively predicted by low status (low S.E.S. and illiteracy/low education of parents) and low occupational characteristics of the family.

Findings of Objective - III

1. A close perusal of the ANOVA results obtained on third objective of the study, revealed that certain psychological dispositions of parents, i.e., attitude towards parents, stress, anxiety and coping responses exereted influence on child abuse and neglect. More specifically, it was found that :

- Maltreating (abusive and neglectful) parents reported negative attitude towards their parents than non-maltreating parents,
- The level of stress was significantly high in maltreating parents than non-maltreating parents,
- The level of anxiety was also significantly higher in maltreating parents than non-maltreating parents,
- Both groups of parents differed significantly on coping responses. Non-maltreating parents often exercised active coping styles. Contrary to this, maladaptive coping style and strategies were adopted by maltreating parents.

2. Correlation results indicated negative relationship between attitudes towards parents and child maltreatment (abuse and neglect). Further, stress and anxiety were found positively correlated with child abuse and neglect. Active and adaptive coping were found negatively related with abuse and neglect, whereas, maladaptive coping responses were found positively correlated with child maltreatment.

3. The SMRA evinced that child abuse and neglect were explained by certain mediating processes. Child abuse was favorably predicted by high stress level in parents and negative attitude towards parents.

Despite this, child neglect was contributed by negative attitude, high stress and poor adaptive coping ability.

Findings of Objective - IV

Results obtained on fourth objective, clearly demonstrated the impact of child abuse and neglect on cognitive, motivational and behavioural processes and personality pattern of children.

A. Cognitive Functioning

A (i) Cognitive functioning, viz. perceptual identification, verbal learning and memory, creativity and intellect were examined as a function of child abuse, residential setting, socio-economic status, age and gender. It was found that :

- Perceptual identification was adversely affected by level of abuse, rural residential setting, low socio-economic status, and younger age. Abused children did poor on perceptual identification than non-abused counterparts,

- Rural children performed inferior on perceptual identification,
- Perceptual identification was worst in low SES children, and
- Children as compared to preadolescents and adolescents scored low on perceptual identification.

(ii) Different groups of children differed significantly onverbal learning and memory span. More specifically :

- The verbal learning and memory span were found inferior in abused than non-abused counterparts,
- Rural children achieved low on verbal learning and memory span, and
- Boys showed superior memory span than girls.

(iii) Results revealed that children differed significantly on creative potentials. It was found that:

- Abused children achieved significantly very low on various domains of creativity than non-abused children.
- Rural children expressed poor creativity than urban children, and
- Boys expressed superior originality than girls.

(iv) It is also apparent from the result that different groups of children varied significantly on intellectual functioning. More, specifically, it was found that :

- Abused children expressed poor intellect than non-abused children,
- Rural children were found intellectually inferior than urban children,
- Low SES children performed inferior on intellectual tasks than middle and high SES counterparts respectively, and
- Children showed poor intellect than pre-adolescents and adolescents.

B. Motivational Process

Further, present findings revealed that child abuse adversely influenced the motivational level of children. It was found that:

- Abused children were found low achievers than non-abused children.
- Urban children expressed high achievement level than rural children.
- Boys were found high achievers than girls.
- Adolescents were found high achievers than pre-adolescents and children subsequently.
- Low SES children expressed poor achievement level than middle and high SES counterparts.

The level of aspiration also varied across different groups. It was found that abused children expressed low aspiration level than non-abused children.

C. Behavioral Process

Social behaviour of children was also influenced by level of abuse, residential setting, age, gender and SES. In other words :

(1) Aggression level differed significantly across various groups of children. It was found that:

- Abused children expressed high aggression than non-abused children;
- Rural children were found more aggressive than urban children;
- Younger children expressed higher level of aggression than pre-adolescents and adolescents respectively;

(2) Helping behaviour in children also varied significantly across groups :

- Abused children expressed diminished level of helping than non-abused children;
- Rural children showed more helping tendencies than urban children; however, gender, age and SES effects were non-significant.

D. Personality Pattern

The personality pattern of children also varied in accordance with level of abuse and age :

- Abused children were found high excitable and impatient, more expedient (disregard, rules), more shy and timid, but less intelligent, less circumspect, artless and sentimental, they also expressed more group dependence, poor decision making ability and self- sufficiency, high tension and frustration than non-abused counterparts. Results, further, suggested that children were found less warmhearted, intelligent, less excitable but more assertive, enthusiastic, expedient, shy and difficult, more artful, guilt prone, group dependent, more uncontrolled or indisciplined, more tense than pre-adolescents and adolescents respectively.

2. Apart from this, correlation results also revealed that child abuse and neglect were found negatively correlated with perceptual identification, verbal learning and memory span, creativity and intellectual functioning.

- Achievements and level of aspirations were also found to be negatively related to child abuse.
- Furthermore, child abuse was found positively related with aggression but negatively with helping behaviour.
- The relationship of child abuse with various personality factors was also determined, which denoted varying patterns of relationships between both factors. Results suggested that personality development of children is adversely affected as a result of abusive and neglectful treatments by parents.

3.Prediction of Psychological Consequences from Child Abuse and Neglect

The psychological consequences of child abuse were predicted on the basis of stepwise multiple regression analysis (SMRA). Results indicated the adverse consequences of child abuse on various psychological processes and personality development.

(A) Prediction of Cognitive Processes

(i) Child Abuse as a Whole and Educational Neglect contributed to poor perceptual identification ability in children;

(ii) Poor verbal organization (clustering) was predicted by Child Physical Abuse (CPA) and Abuse as a Whole;

(iii) Low memory span was also contributed by child educational Neglect, Child Physical Abuse, Psychological Abuse and Abuse as a whole respectively;

(iv) Creative potentials, both verbal and non-verbal were also predicted by child abuse and neglect. Poor verbal fluency and flexibility were predicted by Child Neglect as a whole. Despite this, Psychological abuse contributed negatively to verbal originality. Further, Elaboration for Picture and Title (N&V) was inversely predicted by Child Physical Abuse, Psychological Abuse and Child Neglect. Similarly, Child Physical Abuse, Psychological Abuse, Sexual Abuse and Child Neglect have negative contribution to originality for Picture and Title (N&V). Non-verbal flexibility was also negatively explained by Child Physical Abuse and Sexual Abuse, and

(v) The low intelligence level was predicted by Child Neglect as a whole and Psychological Abuse respectively.

Present results, thus, evinced that child abuse and neglect interrupt the adequate development of cognitive functioning in children.

(B) Prediction of Motivational Processes

It is apparent from the regression result that Child Physical Abuse, Physical Neglect and Neglect as a whole significantly disrupt the proper development of achievement and aspiration levels in children.

(C) Prediction of Behavioural Process

The SMRA result revealed that child abuse and neglect predicted heightened aggression level and diminished helping behaviour in children.

(D) Prediction of Personality Patterns

It is evident from the SMRA results that various forms of child abuse and neglect played damaging role in the development of personality. It was found that:

(i) Factor B (Intelligence) was negatively explained by Child Emotional neglect;

(ii) Factor C (Emotional Stability) was also inversely predicted by Child Physical Abuse (CPA);

(iii) Factor D (Phlegmatic *vs.* Excitable) was favorably predicted by Child Physical Abuse and Psychological Abuse (CSA).

(iv) Factor E (Obedient—Assertive) was contributed by CPA and CSA;

(v) Factor F (Scrious *vs.* Happy) was also contributed by CPA and CSA;

(vi) Factor G (Expedient *vs.* Conscientious) was also predicted by CPA and Abuse as a Whole;

(vii) Factor H (Timid *vs.* Venturesome) was explained by child Educational Neglect;

(viii) Factor I (Vigorous *vs.* Doubting) was contributed by CPA and CSA;

(ix) Factor O (Placid *vs.* Apprehensive) was predicted by CPA and CSA;

(x) Factor Q_2 (Group Dependent *vs.* Self-sufficient) was contributed by CPA and CSA;

(xi) Factor Q_3 (Undisciplined *vs.* Controlled) was explained by CPA and;

(xii) Factor Q_4 (Relaxed *vs.* Tense) was predicted by child physical abuse (CPA).

Present findings, thus revealed that child physical abuse, psychological abuse, sexual abuse and child neglects exerted negative impact on proper development of personality.

On the basis of findings of present study, certain issues have been raised :

1. How for these findings are in consonance with the other research findings?
2. At what extent, the ill consequences of child abuse on cognitive, motivational and behavioural functioning are confirmed and how for residential setting, gender, age and socio-economic status influenced these functionings?
3. At what extent, abusive treatments damage the personality development of children?
4. What are the patterns of linkages between predictors and

incidences of child abuse as well as between abuse and its consequences?

5. How and at what extent ecological, socio-demographic and personal (psychological) factors contributed to child abuse and neglect? and
6. How and at what level child abuse and neglect have contributed to various psychological processes and personality development of children?

An effort, therefore, has been made to discuss these issues in the light of findings of this research as well as other empirical and theoretical evidences.

EMPIRICAL VALIDATION OF THE FINDINGS

1. *Child Maltreatment (Abuse and Neglect) in Relation to Ecological Perspectives*

The finding of present study suggested that numerous ecological factors exercised significant role in the origin of child abuse and neglect. Since Multiple forces determine the phenomenon of child abuse across four ecological levels: in individual (ontogenetic), the family (the micro-system), the community (the exo-system) and the culture (the macro-system) in which both the person and family are embedded. These factors are nested ecologically to each other (Belsky, 1984). Finding of this study evinced that Child abuse is present at alarming threshold in rural and urban societies, in some forms or other. However, a close perusal of results revealed that some forms of abuses were found significantly higher in rural society as compared to urban societies. More specifically, child physical abuse, emotional neglect, educational neglect and neglect as a whole, were found greater in rural families than their urban counterparts, whereas, on psychological and sexual domains, children were found equally abused in urban and rural societies. It appears that in both communities children are abused; only dimensions and forms of maltreatment differ. Since the traditional pattern of child rearing is highly accepted in rural community and therefore, rural families exercise corporal punishment to discipline /control their children, contrary to this, attitude towards child rearing is gradually changing in urban societies. Urban parents are more interactive, co-operative and apply democratic pattern to rear their children. As a matter of fact, physical abuse was found greater in rural community than urban families. Rural parents were also found more neglectful. Since high illiteracy/less education, irresponsibility, lack of awareness and larger number of children (Crowded family) in rural families are some of the root causes of high degree of maltreatments of rural children than urban counterparts. Contrary to this, in urban community, certain other forms of child abuse and victimizations, viz. child labor, homicide, foeticide, induced abortion of girl child, child prostitution etc. are found at alarming threshold. (Pandey, 2005; Baruah, 2003; 2002; Gil, 1971).

2. *Child Maltreatment (Abuse and Neglect) as a Function of Demographic Characteristics of Child, Parents and Family*

Present results evinced that numerous demographic factors related to child, family and parents exercised role in the origin of child abuse and neglect. Findings have been interpreted and discussed in the light of other research evidences.

A. *Child Characteristics*

The findings of present study denoted a clear-cut association between certain demographic characteristics of children and child maltreatment. Other studies also confirm that age and gender are the main demographic characteristics, significantly linked with differential reporting of maltreatment (Pandey, 2005, Sedlak & Broadhurst, 1996).

(*i*) *Child Gender:* Results denoted significant variations on child physical and educational neglects among boys and girls. Girls were found more physically and educationally neglected than boys. There may be numerous causes of girl child neglects. In traditional family structure, male members are the main decision makers on any family issue and females are less empowered even to take decision on very personal issues. Son is more preferred due to numerous, socio-economics, religious and psychological reasons, whereas due to certain inherent costs (economic cost, dowry etc.), except religious (*Kanyadan*) factor, girls are less desired and therefore, they are found more neglected than boys and this practice is higher among rural families.

Another study indicated that except sexual abuse no significant difference between boys and girls was reported on any domain of child abuse (Finkelhor, 1993). Despite this, according to National Incidence Study (NIS) report, girls were 25 per cent more likely to be the victims of any form of abuse than boys. However, 80 per cent of the girls become victims of sexual abuse (Finkelhor, 1993). The dynamics of sexual abuse vary for boys and girls. Finkelhor (1995) reported that boys and girls are more likely to be abused by someone, they trust. Girls are more likely to be sexually abused by male family members, whereas, boys are often abused by male non-family members. In a survey, Mehta (1979) found that most of the kidnapped children were girls in the age range of 3 to 16 years with the motive of selling, begging, sexual gratification, prostitution, family dysfunction, unemployment, extreme poverty and antisocial acts.

(ii) Child's Age: Results denoted that forms and extent of abuse and neglect differed significantly across varying age levels of children. Pre-adolescents and children (6-10 yrs.) were found more physically abused than adolescents. However, sexual abuse was identified maximum among adolescents. Other studies also confirm this finding (Sedlak and Broadhurst, 1996). Since older children can escape easily or retaliate than younger children. Contrary to this, the high incidences of sexual abuse in adolescents have certain reasons. Adolescents living in poor economic and emotionally oppressive environment in the family feel helpless and start seeking sympathy and support out side the family. Particularly girls runaway from home or willingly leave the house with perpetrators, who exploit them for earning purposes and ultimately they become victim of sexual abuse. A study conducted in Delhi revealed that more than half involved cases were minors. Bombay report was sensational which denoted that among 150 cases of rape victims, the youngest was 6-month-old girl.

In a recent event, a girl of 14 Yrs. was lured by an adult (Perpetrator was the father of two children) and willingly she ran away from home with the person to another city. That fellow exploited and sold the girl in Red light area. Luckily, she was saved by police intervention and the man was put into jail (*Times of India*, Dec. 2004). Thus, such interventions are urgently required to check this evil. According to NIS (National Incidence Study-III) report toddlers, pre-schoolers and younger adolescents are the most common victims of physical and emotional abuses, which correspond to the origin of greater independence and parent–child conflicts at these developmental stages. However, child physical abuse exercises influence on a sizeable number of children of all the groups yet, the main reason of physical injury found among older children of 12-17 yrs. is, increasing conflict between parents and teenagers.

(iii) *Child's Education:* Results also denoted a negative association between child education and maltreatment. Illiterate/low educated children were more victims of child maltreatment than highly educated children. Education makes children more aware of their rights and responsibilities. They can rationalize adults and protect themselves from the adult abusive treatments. Educated children can skilfully avoid the circumstances of being victims of abuse and neglect. NIS findings also confirm that low education was positively linked with high rate of child abuse and neglect.

(iv) *Health Status:* Weak children were found more abused and neglected than normal and excellent health children. Contrary to this, handicapped children were more psychologically abused than normal children. Since poor health status of children especially, if health problem is chronic and incurable, parents feel high stress and frustration; consequently, they start maltreating their children. Another reason is, that due to poor health or inability, weak children are unable to fulfil the expectation of parents, teachers or peer groups and significant others, and therefore, are disliked by adults, get less appreciation and acceptance and become victim of abuse and neglect.

B. Family Factors and Child Maltreatment

Results revealed that a number of family characteristics, i.e. family structure (type of family), family size and size of house and family environment were found closely associated with child maltreatment.

(i) *Family Structure:* Results revealed that child maltreatments are found higher in nuclear families than joint families. In fact, nuclear family structure includes variety of burdens, difficulties, and stressors. In a nuclear family, parents are over burdened due to fulfilling domestic and professional interface, as they lack family support system. In case of double career families, it becomes more severe. Other research findings also confirm that females of nuclear families are found more prone to high stress (Pandey and Srivastava, 2004),which is also found closely linked with child maltreatment (Korbin, 1987).

(ii) *Family Size:* Results further indicated that children in larger families were found more abused and neglected than small size families. The crowd in larger family causes feeling of high stress, which in turn contributes to child abuse and neglect. This finding also receives support from NIS research reports (Sedlak and Broadhurst, 1996). They reported that in larger families, children are three times more physically abused and neglected than smaller families. Since additional children in a family means excess responsibility and additional task, so child abuse and neglect are present at high rate in larger families than small size families (Sedlak and Broadhurst, 1996).

(iii) *House Size:* Findings of the study also suggested that different forms of abuse and neglect were found far high in small houses than large houses, since spatial density causes high stress feelings (Stokols, 1972; Pandey, 2002, 2003). Gil, (1971) found that family with high stress feeling were found more abusive and neglectful. Child abuse was mainly caused by stress and frustration resulting from deprivation or poverty, high density as overcrowded and dilapidated housing, large number of children, one parent families, absence of child care alternatives and poor education.

Another characteristic of abusive families is the social isolation of the community coupled with stressful living conditions such as, lack of adequate day care, peer group or close friends and inadequate housing (Thompson, 1994). A close link between family structure and child maltreatment was reported in National Incidence Study. Single parent families were found more abusive than comparison group in which both parents are living together. Apart from demographic characteristics of the family, family environment of abused and non-abused children were differed on various domains. Results revealed that abused and neglected children reported poor levels of cohesion, independence, achievement orientation, intellectual cultural orientation and organization than non-maltreated children. Contrary to this, conflict and control were very high in maltreating families. Korbin (1981) also pointed out that child abuse and neglect were usually observed at high rate in highly disorganized families. Other study also pointed out the role of family dynamics in the incidences of various forms of abuses. National Research Council (1993) observed that range of psychological sequels of abuse often occurs in the center of multifaceted functional factors and poverty, parental psychopathology, alcoholism and family dysfunction are prevalent. Friedrich (1990) also pointed out that family context is generally described as one of the insensitive, marginal parenting and harsh interaction among family members. A sizeable number of studies revealed that distorted family relationship and poor bonds between parent and children give origin to child maltreatment (Hurlock, 1984; Korbin, 1981, Gil, 1971).

C. *Parents'/Adults' Demographic Characteristics and Child Maltreatment*

A close perusal of results revealed that several demographic characteristics of parents/adults were found closely associated with child abuse and neglect. Major findings are interpreted in the light of other empirical/theoretical evidences.

(i) *Parents'/Adults' Age:* Results indicated that younger parents (age below 30 yrs.) were found more abusive than older parents (40+yrs.). Child physical abuse (CPA), Child Sexual Abuse (CSA) and Child Physical Neglect (CPN) were significantly higher among children of younger parents than those of elder parents. National Incidence Study (NIS) also confirms the finding of present study.

Since younger parents lack parenting skills and feel high pressure and stress to cope with daily life hassles, which in turn make them abusive than more experienced and mature parents, who enjoy parenting and have better ability to adapt with stressful situations. Since positive parenting is essential for proper development of children within a safe environment, although, it may cause early helplessness and dependency, but also encourage the child to differentiate and acquire a separate identity. Numerous studies (Steele, 1980; Bowlby, 1969, 73, 80) suggested that development of attachment dynamic is relevant to provide a secure emotional base in early life, which facilitates to the development of self-esteem, a capacity for autonomous functioning and empathy for other. Contrary to this, poor parenting skill and parenting breakdown have positive link with child maltreatment (Belsky, 1984; Wolfe, 1987; Reider *et al.*, 1993).

(ii) *Parent's Gender:* Results revealed that mothers were found more maltreating than fathers on some dimensions of abuse. Child Physical Abuse (CPA), physical neglect and educational neglect were committed more by mothers than fathers. This finding can be explained on the basis of gender based socialization pattern and child rearing practices. In a male dominated society, females project their frustration and unmet expectations on children unconsciously and do abusive treatment with their minors (daughters or sons).

(iii) *Parent's Education:* Findings further reported an inverse relationship between education and child maltreatment. Illiterate or low educated parents were found more maltreating than high-educated parents. Child physical abuse, psychological abuse, abuse as a whole, emotional and educational neglects and neglect as a whole were positively predicted by parents' education. Incidences of child abuse and neglect were found higher among illiterate or low educated parents than highly educated counterparts.

Since educated parents are more change prone and rational and therefore, exercise democratic pattern of child rearing. Contrary to this, illiterate parents believe in corporal punishment as an approved device for disciplining or controlling their children. Due to the lack of awareness and meanness in accepting child dignity they behave in highly abusive and neglectful manner towards their children. Other research findings also confirmed the linkage of poor education level and maltreatments (Sedlak and Broadhurst, 1996).

(iv) *Parent's Health :* It appears from the result that health status of parents was found to be associated with child maltreatment.

Except sexual abuse, all forms of abuse and neglect were committed maximum by handicapped parents than parents with weak, normal and excellent health status respectively. Since chronic illness and physical disability cause acute levels of stress which damages the active coping ability of parents. Such parents also get rejection and comments in society and therefore develop complexes towards own self-concept. Consequently, they project their complexes and frustrations on their children in the form of abusive treatments.

To conclude, numerous demographic characteristics of the child, parents and family contribute to the origin of child abuse and neglect.

(3) Social Status and Occupational Characteristics as Predictors of Child Maltreatment

Child maltreatment was also studied in relation to social status and occupational characteristics of abusive and non-abusive families. Results revealed significant influence of these factors on incidence of child maltreatment. Findings are interpreted and discussed in the light of empirical and theoretical evidences.

(i) Social Status of Families

Since Social status was determined on the basis of caste and socio-economic status of families. Results denoted very high rate of child abuse and neglect in low caste and low socio-economic status families.

(a) *Caste and Community:* Result revealed that various forms of child abuse and neglect, except sexual abuse, were found significantly high in scheduled caste than other backward caste and upper caste of Hindu community. However, maximum incidences of abuse and neglect were found in Muslim as compared to Hindu community. Results are explained on the basis of socio-cultural factors.

Studies indicate that a large segment of population, who are socially disadvantaged or sufferers of social isolation, poverty and social disadvantages are found more abusive. Since social isolation is eventually a noted characteristic feature of abusive families, who lack social recognitions, adequate care, peer groups, close friends and adequate housing (Coulton *et al.* 1995, Thompson, 1994). Abusive parents often lack social control on others in the family, neighbourhood, community and social agencies, which provide needed support and assistance (Korbin, 1994). Numerous studies support that there are multiple factors responsible for high rate of maltreatment in low status families (Straus *et al.*, 1980). Coulton *et al.* (1995) identified that in low class families unstable or violent family situation, several disadvantages (Housing, privacy, noise, pollution, crowding etc.) are very common and emerge as major socio-cultural factors exercising impact on high rate of child maltreatment in North America.

(b) *Socio-economic Status:* Results revealed that children belonging to low SES families were found highly abused and neglected than middle and high SES families respectively. This finding is also confirmed on the basis of numerous other studies, conducted in India and abroad. As National Incidence Study (NIS III) reports, children of low-income strata were significantly high victims of child maltreatment hoverer; children of different strata were also found abused and neglected in some or other forms.

It is essential to note that several factors associated with poverty, may contribute to child abuse and neglect, viz. poor residential setting, low education, high rate of substance abuse and emotional problems. Moreover, families with low socio-economic status have poor social support system to assist/educate parents for childcare responsibilities (Sadlak and Broadhurst, 1996).

(ii) Parental Occupation

Results of this study denoted close linkages between parental occupation and child maltreatments. Maximum abuse and neglects were found in labor class families than cultivators, business and service class respectively. Reasons are obvious and can be explained on the basis of role of numerous socio-cultural and psychological variables. Due to their poor economic status and unstable sources of income, labor class families are bound to live in very pathetic condition. Excessive poverty, illiteracy, social isolation, poor social support and lack of awareness about their rights and responsibilities towards their children, orient parents for harsh treatment. (Kurtz *et al.*, 1993, Trickett *et al.*, 1991).

Results, thus, confirm that low social status and occupations of family positively contributed to abusive and neglectful practices with the children.

4. Psychological Characteristics and Child Maltreatment

Another objective of the study was to explore some psychological characteristics of abusive parents. In present study, parental attitudes, stress, anxiety and coping were investigated and their linkages with different forms of child maltreatment were determined. Findings are interpreted below:

(i) Adult's Abusive Background and Attitude towards Parents

Because of early abusive experiences, parents could develop negative attitude towards their parents. Hence, abusive parents expressed more negative attitude than non-abusive parents. Since prior abuse and family violence in childhood may perpetuate vicious cycle of violence across generation, this issue remains an importment, but not fully resolved matter. Kempe and Helfer (1972) found positive relationship between childhood experiences of abusive treatment and child maltreatment. Other studies also denote that abusive background increases the probability of abusive treatment with their children (Wolfe *et al.*, 1998, Malamuth, *et al.*, 1991).

Certain aspects of child maltreatment are recognized by a sizeable number of researchers. A cursory glance at review of literature on child maltreatment and intervention programmes suggested that:

(a) Child abuse is a widespread phenomenon which cause significant level of harm,
(b) Most adults who were abused as children are aware of all or most salient parts of their abused experiences,
(c) Some adults who were abused as children described that there were periods in their lives when they had no conscious knowledge of their abuse, only to become aware of this in later life,
(d) False beliefs/memories/alligators of child abuse do occur which affect abusers as well as the abused,

(e) Some abusers maintain inexorable stances of denial. For example, abused parents are found as sufferers of false memory syndrome.

Above facts are explored by researchers during the planning of intervention programmes to serve sufferers of abuse (Briere, 1997; Courtois, 1997; Lindsay & Briere, 1997, Read & Lindsay, 1997).

The role of personal/parental factors in child abuse is widely studied during the last three decades. Earlier, 1960s, researchers have focused on a factor known as "Abusive personality type", and suggested that this problem is rooted in adult's psychological disturbance. Abusive adult's were found abused in their own childhood having unrealistic expectations to fulfil unmet emotional needs and aggression control (Spinetta & Rigler, 1972; Kempe *et al.*, 1962), yet latter studies did not fully confirm the role of abusive personality type variable and emphasize certain other parental characteristics, viz. substance abuse, belief in harsh punishment, young age (below 30 yrs) and low educational level, poor parenting skill, mentally disturbed parents, drug abuse and parental stress (Belsky, 1984).

Negative Affects and Child Maltreatment

(i) Parental Stress, Anxiety and Child Maltreatment

It is apparent from the result that abusive parents reported high level of stress than their non-abusive counterparts. A positive relationship between level of stress and child maltreatment (abuse and neglect) was found. Other studies also confirm that abusive parents reported high level of stress than non-abusive counterparts (Korbin, 1994; Gil, 1971). Since stress is a dynamic condition in which an individual is confronted with an opportunity, constraint or demand related to what he or she desires and for which the outcome is perceived to be both uncertain and important. Selye (1956) proposed the concept of General Adaptation Syndrome (GAS) to describe stress when an individual is confronted with a threat, the general psychological response is produced in three stages, i.e., alarm stage, resistance stage and exhaustion stage. However, there are numerous views about stress. The impact of any stressful event is substantially influenced by how a person appraises it. If family environment is extremely stressful and parents/family members are unable to cope with stressful episodes, negative outcome takes place. The finding of present study also denoted a positive association between anxiety and child maltreatment and various forms of abuse were positively explained by anxiety and stress. Further, abusive parents expressed high level of anxiety than non-abusive parents. In another study, Gil (1971) observed that stress and frustration were the major causal factors of child abuse. Investigators have identified exclusive linkages between some negative affects in parents and child maltreatment. They found abusive parents as more impulsive and expressing low frustration tolerance (Averill, 1983; Zillman, 1979).

(ii) Coping with Stress and Child Maltreatment

Present finding evinced that coping contributed differently to child maltreatment. The direction and magnitude of relationship varied among abusive and non-abusive parents. Abusive parents exercised maladaptive styles (Drug usage) to confront with stressful situations and were found infrequent users of active related coping responses than their non-abusive counterparts. Active coping responses were found negatively correlated with child maltreatment. Contrary to this, maladaptive coping was found positively related with maltreatments. A sizeable number of studies indicated that coping (active) is identified a significant moderator in mitigating ill consequences of stress (Pandey & Srivastava, 2004, 2002, 2000; Misra, 1999; Lazarus & Folkman, 1984). Present study evinced that abusive parents failed to exercise active and adaptive coping responses frequently than non-abusive parents, whereas they used drugs (Maladaptive) frequently to confront with stressful episodes. Since active/adaptive coping displays "Buffering role" in stress-illness relationship however, maladaptive coping plays 'Intensing role' in stress-illness linkages. Since abusive parents expressed high level of stress and adopted maladaptive coping (drug use) style consequently, they treated their children in more abusive and neglectful way than their non-abusive counterparts. Findings have been interpreted on the basis of transactional model of stress and coping (Carver *et al.*, 1989; Lazarus & Folkman, 1984).

5. THE PSYCHOLOGICAL CONSEQUENCES OF CHILD ABUSE

Present results evinced the outcomes of child abuse on various psychological processes and personality pattern of children. Findings have been interpreted and discussed in the light of other empirical and theoretical evidences.

(i) Consequences on Cognitive Functioning

The study investigated some of the cognitive functioning, viz. perceptual identification, verbal organization and memory span, creative thinking and intellectual functioning in relation to child abuse, residential setting, gender, socio-economic status and age. Findings are interpreted below.

Perceptual Identification

A close perusal of results evinced that child abuse exercised negative impact on perceptual identification. Hence, abused children were found inferior on perceptual task than non-abused children. Rural setting, low socio-economic status and age also influenced the level of perceptual identification. Poor perceptual identifications were found in rural, low SES and younger children. However, gender difference was non-significant.

A cursory glance at review of literature on perceptual identification indicated the role of contextual/environmental factors on identification of picture. Perceptions of picture are influenced by various socio-cultural backgrounds (Deregowski, 1980; Kennedy, 1974). Apart from this, Witkin's (1962) theory of psychological differentiation suggests that person's psychological differentiation proceeds from less differentiated to more differentiated pattern of functioning, which is manifested in the segregation of his psychological activities (Witkin, Dyk, Faterson, Goodenough and Karp, 1962). The perceptual field differentiation is demonstrated by the disembedding of a smaller geometrical form or a picture from a larger organized background. The level of differentiation is determined by ontogenetic, hormonal and geometric factors (Witkin & Goodenough, 1981). However, ecological demands and cultural pressures have been found to be powerful predictors of the levels of differentiation among individuals

(Berry, 1991; Sinha, 1976). A bulk of studies suggested that individual daily life circumstances and routine activities associated with them tend to have significant impact on development of perceptual identification (Mishra, 1992; Regoff, 1990; Berry, *et al.*, 1988).

Earlier researches (Dyk & Witkin, 1965; Witkin, *et al.* 1962) pointed out the role of a set of socialization variables influencing child's autonomous functioning and his level of differentiation (Witkin & Goodenough, 1981). This ability is influenced by opportunity for encouragement, nature of control over child's aggressive, impulsive and assertive behaviour and parental characteristics, which influenced separation process and in the process of emotional and behavioural control.

Findings of present study indicated that abused children did poor identification than non-abused children. Since abusive and neglectful treatments interrupted the normal opportunity of the development of this ability, as a result, they lagged behind in the process of adaptation to their ecological and cultural contexts and therefore performed low on differentiation task. Thus, role of parental factors is considered very significant in the process of perceptual identification.

Poor performance on differentiation process is also supported by other researchers. Mishra (1996) found that non-scheduled caste, educated and urbanized subjects were found more skillful in perceiving and interpreting pictures than scheduled caste, uneducated and non-urbanized counterparts. In another study, Mishra (1996) showed the effects of daily life activities on children's perceptual differentiation and reported that forest children did superior on perceptual identification than village children. The psychological differentiation among rural-urban children of Nepal was also examined by Niraula and Mishra (2001). They reported that children of urban setting performed better than that of rural children; however, gender difference was non-significant. Results of present study also denoted that younger and low SES children scored low on perceptual identification and these findings are further supported by other researches (Mishra, 2001; Mishra, 1992).

Impact on Verbal Learning

It is apparent from the result that abusive practices with children exercised negative role in the process of verbal organization (clustering) and memory. Abused children performed inferior on verbal clustering and this ability was also found inferior in rural, low SES and younger children. However, gender difference was non-significant on clustering. Correlation result also denoted an inverse relationship between child maltreatment and clustering. Since clustering in verbal learning is influenced by nature of task, frequency, and practice. Misra and Tripathi (1980) evinced that high-deprived (HD) group of subjects performed poor on verbal learning task than low deprived people. Psychologists have interpreted that in free recall; clustering (subjective organization of verbal material) appears and facilitates the amount of recall in case of immediate recall of verbal task (Bousfield, 1953).

Theoretical interpretation of these phenomena can also be made on basis of two assumptions (Tulving, 1966; 1962). Reorganization (clustering) in free recall is the results of habit strength and linguistic preferences earlier acquired by individual and secondly; subjects exercise this to get efficiency in better recall of learned materials.

Another theoretical explanation can be made on the basis of Harlow's (1949-60) principle of "learning to learn". He pointed out that experience (practice) on a particular task, faster the ability of individual to handle new problems. Other studies (Girardau, 1959; Shepard, 1957) demonstrated this effect on children. This theory can be applied in the context of findings of present study. Abused children who are deprived of opportunities and encouragements to interact with variety of stimulus/task, hence, experiential deprivation interrupted their performance level on verbal learning task.

Another reason of low performance by abused children is, their poor/neglected background and conflicting family environment. Such families lack adequate verbal interactions and labeling of objects. Thus, interrupted/inadequate background led to higher threshold for verbal mediation (Jensen, 1965), which in turn, restricts the capacity to solve/ problems by verbal mediation. This assumption has been empirically validated in a variety of studies dealing with linguistic deficiencies (Colvin, 1971), vocabulary (Whiteman & Deutsch, 1968) and verbal concepts (Mickelson & Galloway, 1973).

Influence on Memory

Results denoted that abused children showed low level of memory span than non-abused children. Apart from this, rural children showed low memory span than urban counterparts. Further, boys expressed superior memory span than girls. Correlation results indicated negative relationship of child maltreatment with learning and memory and various forms of child abuse contributed negatively to verbal organization and memory span. Present study gets support from other research findings.

Researches in the field of learning and memory have suggested some strategies concerned with reception, acquiring, coding, organization and retrieval of information in determining retention and recall (Bjorklund, 1988; Ackerman, 1985). In a study Donga (1989) examined the role of some socio-demographic factors, such as age, sex, gender, SES and family size in memory. Researcher showed that child's age, birth order, family size and socio-economic status exercised influence on memory but impact of grade level and gender was non-significant. Dixit (1988) identified variability in memory as a function of strategies used in acquisition and processing of information, whereas; Verma (1988) examined the impact of motivation and personality along with age, sex and sense modality on short-term and long-term memory. Results revealed better memory in older than younger children. Furthermore, memory was superior among high-motivated group and expression was found positively correlated with short-term memory in 12 to 16 yrs. of age children. Thus, the role of ecological variations and other socio-demographic factors in terms of the level of verbal learning and memory, is well documented.

Impact on Creativity in Children

A close perusal of results, obtained on creative potentials, revealed that creativity varied significantly as a function of child abuse, residential setting and gender. On verbal fluency, flexibility, originality and non-verbal elaboration of picture and titles, non-verbal originality for picture and titles and flexibility (non-verbal) were found inferior in abused and rural children, girls as compared to boys expressed low on originality. Creativity, on rest of the domains did not vary significantly among boys and girls.

Results of the study are in congruence with other empirical findings. Pandey (2005) found that creative potentials were found inferior in abused children of different backgrounds. They expressed very few relevant and unreported ideas in verbal fluency, showed low level of flexibility and originality on verbal reasoning task. Similarly, they were found inferior on elaboration of pictures and title as well as originality in picture and title than their non-abused counterparts. It appears that abusive treatment damages the proper development of certain creative potentials in children.

Another finding of this study is, that rural children expressed inferior level of creativity than urban children. Result is confirmed with findings of Lembright and Yamamoto (1965), who showed that urban children performed well on creativity than rural children. Since, urban environment provides better opportunity to interact with enriched environment whereas; rural children are deprived of such contexts essential for the development of creativity. Genderwise variations on expression of originality, are also supported by other researchers (Pandey 2005; 2004; Arasteh, 1968). Torrance (1972) suggested that due to different treatments given by family members, originality differs among boys and girls. Since boys are given more freedom and opportunity to grow independently and take more initiative than girls. Thus, gender-based socialization is one of the reasons of poor expression of originality in girls than boys.

A sizeable number of studies have proved that early peak in creativity is the results of numerous factors such as poor health, pressure, lack of leisure times, family circumstances and poor bond between parent and children (Piers, 1973; Pulski, 1970; Lehman, 1968). Suri (1989) showed that cognitive semantic classes, semantic relation and convergent production of semantic infliction are those factors amounting for rearing for rural children.

Impact on Intellectual Functioning

Results further revealed that intellectual capacity differed significantly across different groups of children. Abused children showed low level of intellect than non-abused children. Apart from this, rural and low SES children scored low on intellectual functioning than their urban and high SES counterparts. Further, child abuse predicted negatively to creative potentials of children.

Numerous studies demonstrated that abused children showed lower intellectual and cognitive functioning than non-abused children on general intellectual functioning and specific measures of verbal learning, memory, verbal language, communication abilities, perceptual motor skills and problem solving (Fantuzzo, 1990; Haskett, 1990; Hoffman, Plotkin & Twentyman, 1984). Other researches also confirm that other forms of abuse and neglect cause deficits in intelligence, language abilities, academic skills and problem solving (Wadarski *et al.*, 1990; Egeland *et al.*, 1983), also abused children showed problems with academic performance and maltreated children were found 2.5 times likely to repeat a grade.

Cognitive functioning may be closely related to the quality of care taken in the home environment. A sizeable number of studies identified a cluster of cognitive deficiencies in sexually abused children (Rust & Troupe, 1991; Einbender & Freidrich, 1989). Dodge, Bates and Pettit (1990) pointed out that sexually abused children showed certain deficiencies, viz.; intellectual delay, academic problem, lower level of overall intelligence, reception and excessive language defects, less creativity and flexibility in problem solving and deficits in language comprehension and verbal abilities. The forms of cognitive deficiencies differ across pre-school, school age children and adolescents. The intellectual ability is influenced by numerous factors. Kumari (1986) found relationship of intellectual commitment and educational interest with intelligence, SES, academic achievement and some personality traits.

Several other studies reported that psychologically abused children perform less well than comparison group on achievement and intelligence task (Scheintuch & Lewin, 1981; Strakey, 1980). Maltreated children from disadvantaged group showed school related problems in ability and academic achievement. Misra and Tripathi (1980) denoted that performance on intellectual task was low among high-deprived subjects than low deprived counterparts.

Findings can be theoretically interpreted on the basis of Piaget (1972) views which assume that the intellectual level of a person reflects both, his genetic potentials and experiential factors. Both factors consistently act together in such a way that it cannot be separated. He stated that development of intelligence proceeds on the basis of sensory, motor and perceptual experiences. Hence, individuals varying in terms of experiential backgrounds, (family environment, Parent- child interaction, Parental treatments) would differ on their intellective performance.

Findings of present study, therefore, could be adequately explained on the basis of socio-cultural environment and experiences. Since cognitive processes develop from indifferent to differential levels, however, environmental/cultural diversity is one of the characteristics exhibited by people worldwide.

Consequences on Motivational Process

Results revealed that abused children expressed low levels of achievement and aspiration than non-abused children. Apart from this, motivation levels varied across setting, age, gender and socio-economic status of respondents. Children belonging to rural setting, young age and low socio-economic status were found low achievers than their comparison groups. Boys were found high achievers than girls. Findings further denoted that the level of aspiration was poor in abused children as compared to non-abused group of children. Child abuse found to be negatively related with achievement motivation and level of aspiration.

These results get support from a sizeable number of studies worldwide. Misra and Tripathi (1980) showed that deprived people were found low achievers than low deprived people. A sizeable number of studies denoted that people from low strata of society have low achievement levels than middle and high SES groups (Angelini, *et al.*, 1970; Uhr, *et al.*, 1969; Nygard, 1969; Srivastava & Tiwari, 1967; Rosen, 1962). In a study, Santrock and Wohlford (1970) found that children of single parent/or whose parents were not alive, were found low on achievement and aspiration levels. The home environment also exercised significant influence on need for achievement in Negro and White culturally disadvantaged boys (De Board, 1969). During the process of development the most important determinant has been found to be the timing of demands (McClelland, 1961). Bartlett and Smith (1968) reported that there are variations on achievement levels in people who were given reward training than those subjects who received punishment training and restrictiveness. It is clear from the results of present study and other research findings, that negative consequences of child abuse on achievement and aspiration level of children are mediated by maltreating experiences during childhood.

(ii) Consequences on Social Behaviour

The findings of this study indicated that child abuse and neglect caused behavioural dysfunctions in children.

Child Maltreatment and Aggression

The results revealed close relationship between child maltreatment and aggression and abuse explained favorably to heightened level of aggression in children. Further, abused children showed high level of aggression than non-abused children. The level of aggression was also found high in rural and younger age children than their urban and pre-adolescents/adolescent counterparts. However, gender and SES differences were non-significant. This result has sound support from a sizeable number of empirical and theoretical evidences.

Aggression is one of the most widely studied areas of the outcome of child maltreatment. A bulk of studies conducted worldwide, confirms the heightened aggression and related externalizing behaviours in children who are victims of physical abuse (Wolfe *et al.*, 1998; Shields, Cicchetti & Ryan, 1994; National Academy of Sciences, 1993). Dodge *et al.* (1997) reported that the experience of child physical abuse included aggressive behaviour and communicate its appropriateness as a problem solving skill. High aggression and other problems in abused children may be related to the witnessing family violence and other negative life events. High level of family conflicts, including verbal and physical aggression has been found to be associated with emotional problems and diminished social functioning (Fantuzzo *et al.*, 1991). Thus, heightened aggressions in abused children are very natural and can be interpreted on the basis of frustration-aggression theory propounded by Dollard and Miller (1950), which suggestes that arousal of aggression, is the outcome of frustration. Children, who are physically abused, gradually become frustrated as being unable to save themselves or retaliate and develop aggressive tendency.

Another finding of present study is that rural as compared to urban children, expressed more aggression. Since, in rural families, physical punishment (maltreatment) is well accepted way of disciplining/controlling the children. Often, other forms of abuse and neglects are frequently found in rural setting. Contrary to this, the attitude towards child rearing is gradually changing in urban societies particularly in high and middle SES families; urban parents therefore, exercise more democratic way to rear their children. They try to understand the difficulties and feelings of their children at various developmental stages. Thus, parental roles might have significant impact in the origin of aggression. The findings of present study are in consonance with other researches. Pandey (2004-a) showed positive association between poor family environment, child abuse and aggressive behaviour in children. In another study, Pandey (2004-b) showed heightened level of aggression in abused children, particularly belonging to low SES group and positive parenting style and parental training exercised favorable impact in preventing child maltreatment and lowering aggression. Fantuzzo (1990) evinced that abused children showed high aggression even after the poverty, family violence, family instability were controlled. It appears that child abuse has independent effect on aggressive behaviour.

The findings of this study are supported by theoretical foundations of aggression. The Frustration-Aggression theory (Dollard & Miller, 1950) suggests that arousal of aggression is the result of frustration. Since, abused children would have developed frustration, which in turn aroused aggressive behaviour. These results are also supported by Berkowitz's (1974) theory of aggression. She added that the environmental condition providing cues of violence or destructions are also responsible in the arousal of aggression. The abusive family environment, media and exposure of violence within the family and neighborhood accelerate aggression. Further, Zimbardo (1970) propounded 'Drive theory' to explain aggression. According to him deindividuation is one of the strongest factors, causing aggression.

As abusive treatments persist for a longer time, abused children start reacting or retaliating with developing age and parents gradually loose their control on children and lack cohesiveness and emotional bond. A vicious cycle of bad parent-child relationship sometimes starts the process of maltreatments consequently; such children develop heightened aggressive tendency and other externalizing problems.

Impact on Helping Behaviour

The findings of present study revealed an inverse relationship between child abuse and helping behaviour and also, helping behaviour was negatively predicted by child abuse and neglect. ANOVA results also denoted that abused children exhibited less helping behaviour than non-abused children. Further, rural children showed higher level of helping tendencies than urban children, however, gender, age and SES did not influence significantly.

Since, helping behaviour is determined by numerous personal, social and situational factors. Abusive families provide poor model to imitate helping behaviour. As a result, abused children reported low level of helping. Apart from this, conflicts and restrictions in the family, interrupted the development of helping attitude in children. In this regard, parenting style and child rearing practices play very important role. Pandey (2004) found an inverse relationship between child abuse and helping behaviour. Further, she reported that parental training played positive impact to inculcate helping behaviour in children. As a result, they showed more helping in post-session (after intervention than) pre-session.

Despite a sizeable number of studies indicated numerous social difficulties, including deficits in pro-social behaviour, as a result of child abuse and neglect. Diminished helping in urban children is very natural. City/urban life are more taxing in terms of time, cost and effort. Urban people suffer from information overload and crowding stress. Hence, they lack interaction and behave less pro-socially than rural people. Apart from this, urban children are more ambitious to fulfil their goal, so they study/work more to accomplish their career. Hence, they find shortage of time to invest with helping others than rural counterparts. A sizeable number of studies revealed that abused children displayed deficiency in pro-social behaviour and communication and showed physical aggression. They also showed disturbed peer interaction, deficits in pro-social behaviour, social withdrawal and isolation (Prino & Peyrot, 1994; Crittenden, 1992; Holfman-Plotkin & Twentyman, 1984).

Present results are also supported by various theories of pro-social behaviour. Since intervention for helping the victims, includes multiple stages of cognitive processes. Bib-Latane and Darley (1970) developed a cognitive model of Bystander intervention. They propose that there are several stages in the process of decision, leading to helping or not. Helping a victim is an emergency and decision for helping at any step is critical. There are numerous ways in which Bystander (Helper) can decide not to help. If person notices the event, defines it as an emergency, accepts personal responsibility, and makes decision about what to do and implements the decision, then victim receives help.

Despite this, Sociobiologists (Trivers, 1971) citing several examples of helping, emphasize the role of biological factors in the development of helping. Contrary to this, some psychologists consider helping as a product of social evolution. Two norms play relevant role in this process, viz., the norm of social responsibility and the norm of reciprocity. Earlier prescribes that we should help them, who are dependent on us, parents are supposed to help children. Norms of reciprocity suggests that we should help those who help us. These norms are inculcated in children by parents and significant others. If family environment fails to develop such norms, children become selfish and show low level of helping behaviour and it happened in case of abused children.

Another perspective of helping emphasizes the importance of learning. Developing children learn their social norms about helping. They are taught at home, at school and in the community by adults. Children learn to help through reinforcement of reward and punishment (Midlarsky *et al.*, 1973) and modelling by observing others (model) who act helpfully (Rushton & Teachman, 1978). Since, abusive families provided poor model to imitate helping than non-abusive families, therefore, abused children express diminished tendency of helping. Empirical and theoretical bases of helping behaviour, thus, largely support the findings of present study.

Impact on Personality Development of Children

In this section, the consequences of child abuse on personality development of children have been interpreted and discussed on the basis of empirical and theoretical evidences. Findings of the study revealed that child abuse caused damaging impact on personality development of children. The positive domains of personality was found to be negatively correlated with child maltreatment whereas; negative domains were found positively associated with child abuse and neglect. Further, different dimensions of personality were significantly predicted by child maltreatments. Results indicated that abused children were found less intelligent, expressed weak ego strength, shyness and group dependency. Apart from this, they were found more excitable, disregarding rules, artless, more tense and frustrated. Contrary to this, non-abused children showed high intelligence, strong ego strength and high patience. They were found more rules bound and circumspect (more practical) and therefore, they showed better leadership capacity, decision-making ability and self-sufficience, however, were found less tense and frustrated.

Finding of the present study thus, suggests that abusive practices by parents and significant others, adversely affected the proper development of personality. Consequently, such children lag behind in several other fields, viz.; in education, social interaction, and adjustment within and outside the family. They develop deficits in decision-making and other cognitive functioning to deal with difficult task and adjustment with challenging environment. High tension and frustration may cause mental illness and high aggressive tendency. Since abused children disregard rules and therefore, may be involved with anti-social activities. Further more, poor ego strength and high sensitivity; impractical tendency of abused children would cause difficulties in adjustment with others, in the family, community, and society.

The findings of present study are in consonance with other research evidences. Fisher *et al.* (1997) showed many personality disorders in physically abused children. High rates of depression and conduct disorder were found in abused adolescents (Pelcowitz *et al.*, 1994). Child sexual abuse causes severe long-term and short-term consequences on personality development. They showed low self-esteem, guilt and shame (Conte & Schuerman, 1987; De Francis, 1969). Other studies report high rates of depression, withdrawal and suicidal tendencies (Gladstone, *et al.*, 1999; Kolko *et al.*, 1988; Friedrich *et al.*, 1986). Researches, observed more common symptoms of fearfulness in sexually abused children (Browne & Finkelhor, 1986). A sizeable number of studies indicated that psychologically maltreated children showed lower level of social competence and adjustment problems and high level of insecurity (Egeland, 1991; Vissing *et al.* 1991). Other studies also confirmed the presence of low self-esteem, anxiety, depression, dissociation and interpersonal sensitivity in adults who were psychologically abused as a child (Downs & Miller, 1998; Kent and Waller, 1998; Briere & Runtz, 1988, 90).

Since development of personality is determined by maturational and environmental factors and these factors have sound impact on personality development. Thus, unfavorable family environment, poor parent-child relationship, poor parenting skills, unfavorable neighbourhood and school environment and inputs from community etc. positively contributed to the origin of child maltreatment. Since, abusive families have restrictive and conflicting environment and they face isolation from community. Abusive parents often lack social control in the family/community, the neighbourhood and social agencies, which provide needed assistance (Korbin, 1994).

Present findings thus, evince that child abuse and neglect are predicted by multiple contextual, demographic, social and psychological characteristics of abusers and victims of abuse which in turn exercise adverse impact on proper development of cognitive, motivational and behavioural processes and personality patterns of children.

CONCLUSION

Findings of this study, clearly stressed the influence of ecological, demographic, social and psychological characteristics on the child abuse and neglect. Furthermore, results evinced the ill consequences of child abuse on cognitive, motivational and behavioural functioning, and personality pattern of children. Findings have been confirmed on the basis of a huge amount of empirical studies and theoretical evidences. Apart from this, the study also requires theoretical interpretations to achieve sound conceptual base to understand the phenomenon of child abuse and its consequences.

Theoretical Interpretation of the findings Maltreatment (Abuse & Neglect)

Researchers have conceptualized child abuse and neglect as a social-psychological phenomenon embedded in the dynamics of family, social and cultural contexts and have multiple causative roots. To analyze the origin of child abuse, numerous macro level (sociological) and micro level (psychological) theories have been propounded. Both types of theories support the findings of present study.

1. *Sociological Perspectives of Child Maltreatment*

The findings of the present study are in consonance with the theoretical explanations given by many macro level theories, viz.; cultural explanations, structural characteristics of the family, strain theory, social bonding and deterrence theory.

(i) Cultural Explanations

Present result revealed that different forms of child abuse and neglect were reported at varied proportions in different strata of society. The findings are supported by Cultural Explanation approach, which suggests that violence is an accepted and encouraged form of cultural expression in our society. The acceptance of violence in the culture contributes to criminal violence rates in the family or in the society (Tolan & Guerra, 1998; Straus, 1977). The aggression and violence in home reflects the society's tolerance for aggression (Wyatt, 1994). Apart from this, the family norms also influence at certain extent of family aggression. Discipline, from social scientists perspective, is the cost and implantation of cost that parents impose on their children can vary from physical assault to psychological cost and costs also range from legitimate to illegitimate dimensions.

(ii) The Structural Characteristics of the Family

Family structure sometimes makes individuals/families prone to violence (Brinkerhoff & Lupri, 1988). The pattern of interaction with family members, which causes tension and emotional insecurity, intensity, make family relationship volatile. Another factor is the power differential between family members. Children, subordinate to parents, often become target of violence. The privacy of family also facilitate child maltreatment by making violence relatively easy to hide and provide autonomy in deciding how to discipline children? Family environment is another factor, found to be exclusively associated with child abuse and neglect.

(iii) Strain Theory

Present result also denoted that child abuse and neglect were found maximum in low caste (SC/ST), low socio-economic status and low occupation status families. This finding receives support from "Strain Theory", which suggests that high rates of child maltreatments are found in low income families, unemployed families and families receiving aids (Limber & Nation, 1998). The unequal distribution of opportunities along with the stressors associated with poverty (financial worries, ill health and crowded living conditions), produces high levels of stress and frustration in lower income families (Farrington, 1980; Straus, 1980). Pandey (2004) reported positive association between child abuse and parental stress. As stress and frustration become overwhelming, adults' aggression may be directed to innocent children in the form of abusive acts. The structure of violence theory suggests that the lower class violence to some extent is a cultural norm, a way of life (Wofgang & Ferracuti, 1972).

(iv) *Social Bonding Theory*

Another finding of the present study is that parents with certain demographic and socio-psychological characteristics were found highly abusive and this result can be explained on the basis of "Social

Bonding Theory". The theory attempts to answer the questions that in a given cultural and social context, which factors encourage family violence? Why some people are more violent than others? Since, Social Bonding Theory proposes that most people are sufficiently motivated to commit deviant act (child abuse and neglect) yet what social forces prevent certain people from committing abusive acts? People do not commit deviant acts when they have a stake in conformity. Three components of Social Bonds determine the possibility of child maltreatment, viz.; Commitment, Attachment and Belief (Travis, 1993; Hirschi, 1969). Present results, along with numerous researches on child maltreatment are also supported by Social bonding theory. Young, single parents typically have less stake conformity along with higher rates of child maltreatment (Gelles, 1989). Socio-economic status also contributes to child maltreatment through its impact on social integration (attachment bond). Families who are highly involved in the community life are less likely to be maltreating.

In the present study, young, divorced/separated disabled/poor health parents, belonging to low status (caste and economic status) were found highly abusive than comparison groups. These findings can be interpreted on the basis of social bonding perspective. Parents/families who lack important attachments have less of a stake in conformity and therefore, more likely to be abusive. Similarly, families with positive views of the community and active in community are less likely to be violent (Garbarino & Kostelny, 1992).

(v) *Deterrence Theory*

The findings of present study are also supported by the Deterrence Theory (Gibbs, 1975) also known as Control theory, which focuses as the influence of formal social controls. Gelles and Straus (1988) suggest that parents abuse their children when there are insufficient legal structures to keep them from loosing control. Straus *et al.* (1980) have pointed out that child maltreatment often takes place "Behind Closed Doors", where physical separation from society and family privacy provide parents considerable autonomy. Therefore, the cultural acceptances of parent-to-child violence blur the line between legitimate discipline and child maltreatment. However, the mandatory reporting laws and public awareness are increasing pressure to save children from abusive families and to prosecute abusive parents (Besharov, 1996).

2. *Psychological Perspectives of Child Abuse and Neglect*

Although sociological theories successfully explain why child abuse exists in a society but they fail to explain why individual family members are violent? Primarily, these theories interpret child abuse in structural terms, but do not explain variation in individual's cognitions and behaviours. Explaining the behaviour of individual is more difficult than explaining structural pattern of maltreatment, hence certain psychological theories have made effort to identify risk factors that make maltreatment more understandable and predictable. Thus, interpretation of present results, in relation to psychological perspectives has also been made in nutshell.

(i) Biological Factors

Present result and other studies also support that biological factors of abusers are found to be associated with child abuse and neglect. Abusers show hyper responsive physiological activity to both positive and negative child stimuli. Such activity might make the abuser more physiologically reactive in stressful situations with children and lead to abuse (Milner & Chilamkurti, 1991). Other biological factors, i.e. I.Q. neuropsychological deficiencies, deficit disorders, physical disabilities and disorder and health problems were found closely linked with child maltreatment (Pandey, 2004; Milner, 1998; Crittenden, 1992).

(ii) Psychopathology

One popular explanation of child maltreatment is that mentally disturbed individuals commit more abusive or deviant acts (Kempe, *et al.*, 1962). Hunter *et al.* (1990) reported that higher rates of psychological disorders were found in offenders of child maltreatment. However, some researches argue that psychopathology is found only in a few individuals who physically or sexually abused their children (Milner, 1998; Williams & Finkelhor, 1990).

(iii) Psychological Characteristics

The findings of the present study denoted that abusive parents expressed high levels of stress, anxiety and negative attitudes than comparison groups. In addition, these parents exercised more maladaptive coping (drug use) responses to confront with stressful episodes. A sizeable number of researches also identified that maltreating adults showed higher level of cognitive distortions, depressive symptoms, stress, low self-esteem, more substance abuse or dependence as compared with non-maltreating adults (Milner *et al.*, 1995; Ethier *et al.*, 1995., Chan, 1994). Certain other studies reported that abusive parents display more anger control problems, hostility and low frustration tolerance than non-abusive parents (Milner *et al.*, 1995; Bauer & Twentyman, 1985). It appears that certain psychological and behavioural factors lead to child maltreatment.

(iv) Interaction Theories

The results of present study are also confirmed by the reciprocal nature of the parent-child relationship. This theory suggests that difficult child behaviour interact with behavioural problems and deficits of parents, viz.; poor parenting skill or anger control problems (Pandey, 2004; Milner, 1993). Hollowitz and Sulberg (1959) presented a vicious cycle of bad parent child relationship, which is resulted in the form of child abuse (cited by Hurlock, 1984). Since, abusive and neglecting families are characterized by disturbed patterns of infant's attachments (Kolko, 1992).

Attachment Theory suggests that attachment is an enduring emotional bond that develops during the first year of infant's life. Failure to form a secure attachment in early life may contribute to problems in adulthood in the form of inability to develop close personal relationship (Ainsworth & Bowlby, 1991; Frankel & Bates, 1990). The insecure pattern of attachment was found in abused, neglected and psychologically abused children (Pandey, 2004; Crittenden, 1998). Researchers stressed that insecure attachments are caused by parental unresponsiveness to the physical and psychological needs of the Child. Poor bond between parent-child relations may lead to difficult child behaviours and increased challenges for parents as the child develops, resulting in an inverse escalation of abusive parent-child interactions (Crittenden, 1998; Cerezo, 1997; Hurlock, 1984).

(v) *Role of Social Learning Theory*

Findings of the study have also been successfully analyzed on the basis of Albert Bandura's Social Learning Theory (1971). He emphasized the process of modelling in which individual learns social and cognitive behaviours by simply observing and imitating model's behaviour. Children are exposed to violence directly or indirectly with a set of norms and rationalizations that justify violence. Children acquire that violence is an acceptable way of resolving family conflicts and expressing emotions. Thus, observation and reinforcement of violence within a social context teach children how to perform abusive acts.

The application of this theory has achieved empirical validation on several points. First, violence tends to perpetuate itself from one generation to the next, "like father, like son", "like mother, like daughter" (Straus *et al.*, 1980). Secondly, numerous laboratory experiments with school children evinced that aggression can be learned through modelling (Bandura, Ross & Ross, 1961). Lastly, a sizeable number of studies have linked the exposure to violence in childhood to violence in adulthood (Cappell & Heiner, 1990; Williams & Finkelhor, 1990; Widom, 1989). Although, intergenerational patterns of child maltreatment is criticized and requires further empirical validations (Widom, 1989; Kaufman & Zigler, 1987).

Further Evaluation of Current Theoretical Status of Child Maltreatment

In the light of the present study and other researches, some issues have been raised, which are unresolved in these theories.

The macro level sociological theories explain very well the structural roots of child maltreatment. The cultural explanations focus on broad cultural forces that may promote abusive activities whereas, structural characteristics of the family suggest the demographic characteristics of that family which contribute to child maltreatment. Similarly, sociological theories of deviance, applied to explain child maltreatment are : Strain theory, Social bonding theory and Deterrence theory. Although these sociological theories successfully discuss why child maltreatment exists in a society or why some segment of society experience more violence than others but they fail to explain the process involved in the origin, maintenance and consequences of child maltreatment. Since explaining the abusive behaviour of individuals, is more complex than explaining structural patterns, consequently the answer becomes difficult. Further, Macro level theories cannot properly explain why some adults mistreat their children more and others do not? And, therefore, it becomes difficult to predict who will be violent?

Thus, micro level psychological theories, viz.; Psychopathology, personality traits, biological factors, interaction theories and social learning theory have been applied to explain the processes involved in abusive behaviour. However, biological, psychopathology and personality traits mainly emphasize the root causes of problems within the person, whereas, the social learning theory and the interaction theories focus on major psychological processes in relation to social context (viz.; family and social agents) in the origin of child maltreatment.

Apart from this, there is a need to develop a wide range theory of child abuse, which can better explain, what is child abuse? How is child abuse? And why is child abuse? Hence, it is essential to emphasize the role of mediating and moderating factors in between the antecedent and consequences of child abuse. Present researcher endeavored to incorporate these factors in the relationships of antecedents and consequences of child abuse.

On the basis of present findings, a model has been depicted to show the dynamics of child abuse and its consequences and to discuss the relationships between contextual/environmental, mediating and outcomes variables (Fig. 4.1)

First level of the model contains environmental and social factors. The environmental features include: residential setting, size of house/ living space, neighborhood and family environment and the social factors involve socio-economic status, income, caste/community and occupational characteristics of parents/family. These factors have clear-cut linkage with child maltreatment (abuse and neglect) and its consequences.

Second level of the model includes the demographic and psychological characteristics. The demographic features include: Child's age, gender, education, health and birth order; Parent's age, gender, health status, marital status and education, and family characteristics, i.e., type of family, number of family members, number of male/female children etc., whereas psychological characteristics involve parents, adults, abusive experiences or negative attitude towards their parents, anxiety, stress and poor coping ability in parents (perpetrators). In the dynamics of family, the child has focal place confronting any sort of interaction with parents or adults, who may be found in perpetrator or protector roles and the nature of interaction determines the frequency and intensity of child maltreatment.

Third level of the model indicates status of child abuse and neglect which are originated and shaped by environmental/contextual, demographic and psychological factors and interrelationships of these factors. Within and outside the family, child abuse/neglect may take place as a whole or in any forms of Child physical abuse (CPA), Child psychological abuse (C Psy. A), Child sexual abuse (CSA) and Child neglect, viz., physical, emotional and educational neglects, which in turn exercise damaging impact on psychological development of children. As Fig. 4.1 indicates child abuse along with interrelationships of contextual and personal factors has damaging consequences on development of cognitive, motivational and behavioural functioning as well as personality patterns of children.

The findings of present investigation evince that abused children display distorted perceptual identification, poor verbal organization, low memory span, inferior creative potentials and low intelligence level. It retards achievement and aspiration level of children and also causes behavioural problems, often abused children express heightened aggression and diminished helping tendencies. It also disrupts proper development of personality. Present results reveal that abused children have displayed low intelligence level, high impatience and have shown tendency to disregard rules, weak super ego and low vigor. They are found shy, timid, impractical, and poor decision makers, however, they express high levels of tension and frustration.

The findings of this research investigation have strong support with a sizeable number of empirical and theoretical evidences. It has also close link with Lewin's principle of life space [B = f (P&E)], which denotes that any action is the function of interaction between person and environment.

Implication of the Present Study

The present piece of research is not merely a theoretical exercise, but also is based on empirical data from a large segment of population, belonging to different strata of society in eastern districts of Uttar Pradesh. It is an intensive family/community based study; hence, obtained results have both the theoretical and practical implications.

This study has identified certain contextual, socio-demographic and psychological factors contributing to child maltreatment, which in turn exercised damaging impact on cognitive, motivational, behavioural and personality development of children. More

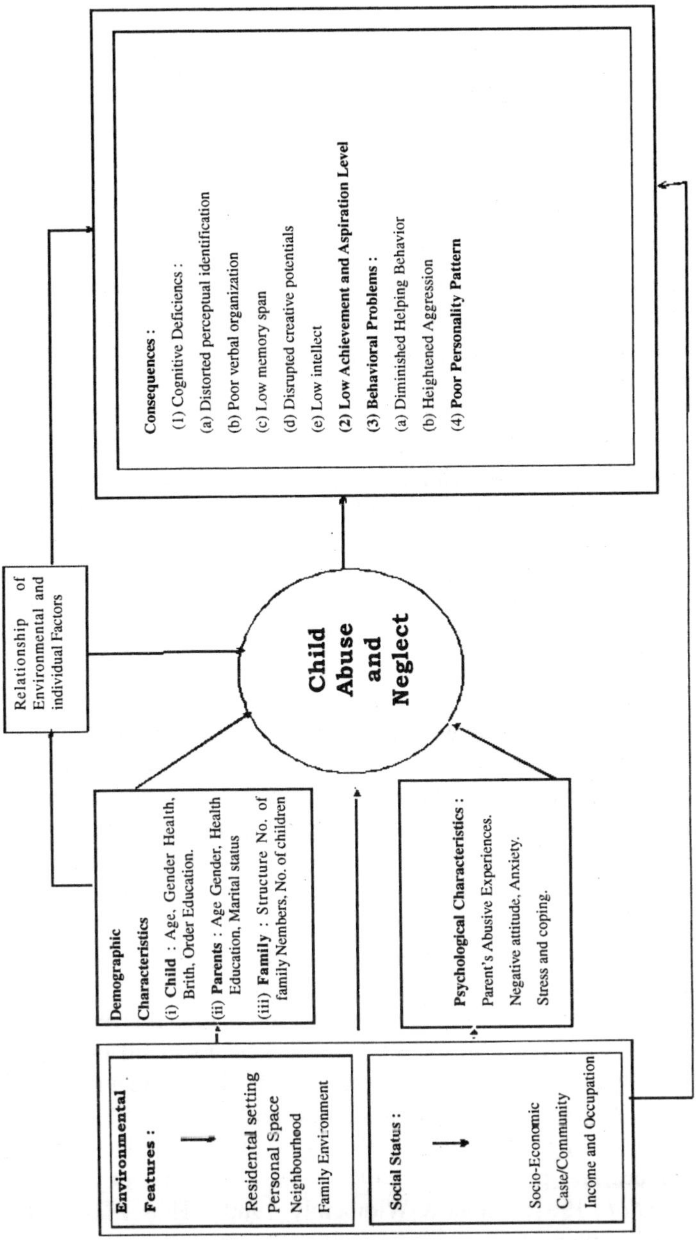

Fig. 4.1 : Dynamics of Child Abuse and Its Consequences

specifically, families belonging to rural residential setting and low status, (i.e., caste, socio-economic-status and occupation), were found more abusive than their comparison groups. Certain personal factors were also found as predictors of child maltreatment. Results indicated that parents with young age (below 30 yrs.), low education level, poor health or disability were found more abusive. Further, parents expressing high levels of stress, anxiety, negative attitudes and maladaptive coping ability were found more abusive. Apart from this, certain child characteristics were also found closely linked with child maltreatment. Greater number of victims of child abuse and neglect were girls, younger children, illiterate or low educated, poor and weak/ disabled children. It is also apparent from the findings of present study that abused children expressed high level of cognitive deficiencies. They were also found low achievers and showed low level of aspiration. Similarly, heightened level of aggression and low helping behaviour were observed in abused children. Apart from this, they expressed deficiencies on certain dimensions of personality. Hence, a theoretical framework was constructed to show the dynamics of child abuse and its psychological consequences (Fig. 4.1).

Study, thus, suggests that abusive parents as well as abused children should be considered most vulnerable/high risk group. They should be exposed to innovative intervention programmes, separately. The results indicate that child abuse is present in eastern districts of U.P. at alarming threshhold within and outside the family. Thus, factors contributing to child abuse and neglect should be systematically chalked out. Innovative intervention programmes, focusing to family or community, or high risk/vulnerable groups, should be developed and implemented at many points of primary prevention (to reduce the occurrence of new cases), secondary prevention (to reduce the duration and severity of prevailed cases) and tertiary preventions (rehabilitation, changing attitude towards abusers and abused). This effort will help reduce the various forms of deficiencies and will promote the quality of life. Despite all the preventive efforts at international levels, child maltreatment continues to be a pervasive Psycho-social problem. Epidemiological studies denote that millions of children are maltreated every year. A sizeable number of studies have demonstrated direct linkages between parental stress, parenting skills, knowledge of child development and potential for child maltreatment (Pandey, 2004; Trickett, *et al.*, 1991; Milner & Wimberely, 1980). Thus, in order to design and implement effective prevention programmes, a thorough understanding of the nature and root causes of child abuse and neglect, is barely essential. Parents, abused as children, with high stress experiences and negative attitude were found highly abusive, hence they need training for positive parenting (Pandey, 2004). Treatment programmes for abused children (Connor, 1987). Stress management and skill training (Egan, 1983) have also been successfully exercised by researchers.

The micro level ecological factors studied in present study are: family environment, dysfunctional relationship between family members and poor marital relationships. Certain intervention programmes at micro levels have been implemented successfully to mobilize favorable family environment. Some of these programmes are: Parenting skills training (Pandey, 2004; Wolfe & Manion, 1984), Educational programmes for parenthood (Zigler, 1980), Marital counselling (Lutzker, *et al.*, 1984) and Enhancement of parent-child interaction (O'Connor, *et al.*, 1980).

Findings clearly indicated that there are great possibilities of enhancing the competencies and personality pattern of children, through preventive and promotive intervention programmes. However, success of preventive programmes is contingent upon the systematic identification of target groups, requirements and feasibility of intervention programmes. Study, thus, suggests that any training, remedial or educational programme should be organized to abused group of children belonging to different backgrounds. A sizeable number of researchers on different samples, demonstrated the positive impact of intervention programmes on enhancement of cognitive abilities (Pandey, 2004; Wilkinson & Murphy, 1976). Results are also supported by some other studies. Sinha (1976) reported that the education children received, contributed in reducing the performance differentials between high and low caste groups of children. McClelland and Winter (1969) showed that direct education method changed the motivation level of children. Pandey (2004) found that children showed improved intellect, low aggression and high pro-social behaviour after attending intervention programmes.

Since, nature and problems of child abuse and neglect are somewhat different in India than those of western countries in many ways, due to the contextual/ecological or cultural variations. In India majority of population belongs to rural community and except a few features, there are no remarkable variations between rural and urban life style in eastern districts of U.P. Child physical abuse (CPA) is an accepted way to discipline the children in these families. Apart from this, in both settings, other forms of abuses are also found in plenty. Thus, the dynamics of child abuse and neglect have to be reanalyzed and tackled in different manner in such diverse settings. The research strategy should also be developed according to Indian context. In our country, poverty, deprivation and population density are the pervasive causal factors of child maltreatment. Thus, preventive intervention programmes should be targeted to vulnerable groups to control the consequence of child maltreatment and save the children from its damaging outcomes. Findings of the study, therefore, suggest the need of preventive and promotive interventions focusing to high-risk group (abusive families as well as the maltreated children), the community and larger society.

Suggestions for the Further Study

A critical analysis of the findings of the present study indicated some of the limitations of this research, which led to fresh views for further investigation.

(1) In the present study, child abuse and neglect were studied within the family but it was realized that children directly or indirectly experience different forms of abusive treatments from outside the family by school-teachers, neighbours, peer groups, relatives and/or any adults, significant to the child. Thus, these groups should also be incorporated in the list of perpetraters of child maltreatment in further researches.

(2) Present study has mainly focused on ecological, socio-demographic and personal factors as root causes of child maltreatment. However, it is also essential to highlight the role of genetic factors in the origin and maintenance of child abuse and neglect.

(3) This study has some methodological limitations and this issue has continuously been raised worldwide by a sizeable number of researchers who are intensively working in this area. In this study also, it was realized during data collection, that respondents tried to fake or hide the correct reporting of abusive activities. They twisted their responses on structured measures. So, there is need to apply multiple methods i.e., in addition to structured checklist or interview schedules,

indirect (unstructured methods) methods, case study and observations should also be exercised to identify the frequency and intensity of child maltreatment. Since child maltreatment is a sensitive issue of research and therefore, social desirability at some extent would have interrupted the correct reporting, therefore, multiple methods should be exercised for the accuracy of data and assessment of this problem. These considerations will be taken in further researches.

(4) Since child abuse is a psycho-social phenomenon and dimensions of abuse vary from culture to culture. This issue is extensively studied in western countries and the definition and conceptualization of child abuse and neglect are primarily, formulated by western researchers, where child maltreatment is a widely studied area. Thus, there is a need to provide more comprehensive definition of child abuse, emphasizing cross-cultural perspectives. Thus, conceptualization of child maltreatment will be done in a more indigenized manner in future researches.

(5) It is also true that some forms of abuses are culture specific whereas certain forms of abuses are universally common throughout the world. However, in present study the measures of abuse and neglect have been developed mainly on the basis of universal forms of child abuse and neglect. But it is realized that some specific forms of abuse and neglect should also be included with the items of abuse checklist and neglect scale. This precaution will be taken in future researches.

(6) Although, certain forms of abuse and neglect were reported by adults (parents) on open-ended interview schedule. It was observed that some forms of child maltreatments are community specific. For example, in poor community drug abuse, sexual harassment, physical punishment, family conflicts and violence are found very high whereas; in middle and high SES families, psychological/emotional abuses are found at alarming threshhold, thus, these groups should be made aware of the consequences of abusive activities through preventive interventions. This study provides an indication for further research.

(7) In the present study, the impact of child maltreatment on cognitive functions has been thoroughly investigated but many other aspects of motivational and behavioural processes have

been less emphasized. Therefore, to understand the influence of abusive treatments, other forms of motivational and behaviour processes in relation to abuse and neglect, should also be investigated in further researches.

(8) In this study, a huge amount of variables were assessed on representative sample. Statistical treatments were made in terms of ANOVA, Correlation and Stepwise Multiple Regression Analysis (SMRA). However, several other forms of fine analyses (i.e., Factor Analysis, Multiple Analyses of Variance etc.) should also be done for better understanding of the concept and to make comparisons at various levels. This effort should be made in further study.

(9) Present result clearly suggests that incidence of child abuse and neglect are present at an epidemic rate in Eastern Districts of Uttar Pradesh and damaging the short and long-term developments of a sizeable number of child population. Thus, preventive efforts are needed urgently. Curative as well as preventive approaches should be exercised. Intervention programmes should be designed for preventive and promotive purposes. Thus, in further researches, such preventive interventions would be designed and implemented to improve the status and quality of life of the children as well as the whole community.

Bibliography

Aber, J.L. & Allen, J.P. (1987). Effect of maltreatment on young children's socio-emotional development: An attachment theory perspective. *Developmental Psychology,* 23 (3) 406-414.

Aber, J.L. & Cicchetti, D. (1984). The socio-emotional development of maltreated children: An empirical and theoretical analysis. In H. Fitzgerald, B. Lester, & M. Yogman, (Eds). *Theory and Research in Behavioral Pediatrics.* (Vol. 2, pp.147-205). New York: Plenum.

Aber, J.L., Allen J. P., Carlson, V. & Cicchetti, D. (1989). The effects of maltreatment on development during early childhood: Recent studies and their theoretical, clinical and policy implications. In D. Cicchetti & V. Carlson (Eds), *Child Maltreatment: Theory and Research on the Causes and Consequences of Child Abuse and Neglect,* (pp. 597-616) Cambridge, U.K. Cambridge University Press.

Ackerman, B.P. (1985). Children's retrieval deficit. In C.J. Brainerd and M. Pressley (Eds). *Basic Processes in Memory Development: Progress in Cognitive Development Research.* New York: Springer: Verlag.

Adams-Tucker, C. (1981). Proximate effects of sexual abuse in children: A report on 28 children. *American Journal of Psychiatry,* 139,1252-1256.

Agarwal, A. & Tripathi, L.B. (1980). Time Perspective in achievement Motivation. *Psychologia,* 23, 50-62.

Ainsworth, M.D.S., & Bowlby, J. (1991). An ethological approach to personality development. *American Psychologist,* 331-341.

Alessandri, S.M. (1991). Play and social behavior in maltreated preschoolers. *Development and Psychopathology,* 3, 191-205.

American Professional Society on the Abuse of Children (APSAC), (1995). Guidelines for the psychosocial evaluation of suspected psychological maltreatment in children and adolescents. Chicago: Author.

Anderson, C.A., Deuser, W.E. & Deneve, K.M. (1995). Hot temperature, hostile affect, hostile cognition and arousal. Tests of general model of affective aggression. *Personality and Social Psychology Bulletin*, 21, 434-448.

Angelini, A.L., Bitencourt, L., Jose, F. & Rosamilha, N.C., (1970). Motive de realizacao edeseuvolvimento economico. *Reuisto Intermericanade Psychologia*, 4, 33-41.

Arasteh, J.D., (1968). Creativity and related processes in the young child: A review of the literature. *Journal of Genetic Psychology*, 112, 77-108.

Atkinson, J.W., (1964). *An Introduction to Motivation*. Pinceton, NJ: Van Nostrand.

Atkinson, J.W., (1983). The achievement motive and recall of interrupted and completed tasks. *Journal of Experimental Psychology*, 46, 381-390.

Augoustions, M. (1987). Development effect of child abuse: Recent findings. *Child Abuse & Neglect,* 11 (1), 15-27.

Averill, J.R. (1983). Studies on anger and aggression. Implications for theories of emotion. *American Psychologist*, 38, 1145-1160.

Azar, S.T. (1997). A cognitive behavioral approach to understanding and treating parents who physically abuse their children. In D. Wolfe, R.J. Mc Mahon, & R.D. Peters (Eds.). *Child Abuse: New Direction in Prevention and Treatment Across the Life Span*, (Vol. 4, pp. 78-101), Thousand Oaks, C.A.; Sage.

Bandura, A. (1971). *Social Learning Theory*: Morris Town, NJ; General Learning.

Bandura, A., Ross, D., & Ross, S.A. (1969). Transmission of aggression through irritation of aggressive model. *Journal of Abnormal and Social Psychology*, 67, 575-582.

Bartlett, E.W. & Smith, F.P., (1968). Child rearing practices, birth order and the development of achievement related motives. In Evans, E.D. (Ed.) *Children: Reading in Behavior and Development*. N.Y.: Holt.

Baruah, A. (2003). *Child Abuse*. References Press. New Delhi.

Bauer, W.D. & Twentyman, C.T. (1985). Abusing, neglectful and comparison mothers responses to child related and non-child related stressors. *Journal of Consulting and Clinical Psychology*, 53, 335-343.

Beeghly, M. & Cicchetti, D. (1994). Child Maltreatment, attachment and the self-system. Emergence of an internal state lexicon in toddlers at high social risk. *Development and Psychopathology*, 6, 5-30.

Behera, M. (1989). Conservation ability of children: Effects of home environment and caste, M. Phil. Psy. Utkal University.

Beitchman, J. H., Zucker, K. J., Hood, J.E., De Costa, G.A. & Akman, D. (1991). A review of the short terms effects of child sexual abuse. *Child Abuse & Neglect*, 15, 537-556.

Beitchman, J.H., Zucker, K. J., Hood, J. E., De Costa, G. A., Akman, D., & Cassavia, E. (1992). A review of the long-term effects of child sexual abuse. *Child Abuse & Neglect*, 16, 101-118.

Belsky, J. (1980). Child maltreatment: An ecological integration. *American Psychologist*, 35, 320-335.

Belsky, J. (1984). The determinants of parenting: A processes model. *Child Development*, 55, 83-96.

Berkowitz, L. (1974). Some determinants of impulsive aggression: The role of mediated association with reinforcement for aggression. *Psychological Review*, 81, 165-176.

Berkowitz, L. (1983). Aversively stimulated aggression: Some parallels and differences in research with animals and humans. *American Psychologist*, 38, 1135-1144.

Bernstein, D.P., Fink, L., Handelsman, L., Foote, J., Lovejay, M., Wengel, K., Sopareto, E. & Ruggiero, J. (1994). Initial reliability and validity of new retrospective measure of child abuse and neglect. *American Journal of Psychiatry*. 151, 1132-1136.

Berry, J.W., Irvine, S.H., & Hunt, E.G. (Eds.), (1988). *Indigenous Cognition: Functioning in Cultural Context.* Boston. Martinus Nijhoff.

Berry, J.W. (1976). *Human Ecology and Cognitive Style*. N.Y., Wiley.

Berry, M. (1991). The assessment of imminence of risk of placement: Lessons from a family preservation program. *Children and Youth Services Review*, 13, 239-256

Besharov, D. (1996, December, 1). When home is hell: We are too reluctant to take children from bad parents. Washington Post, pp. C1, C5-C6.

Bhattacharya, A.K. (1979). Child in India and nutritionally battered child, in R.T. Helfer and C.H. Kempe (Eds). *Child Abuse and Neglect*. Great Britain: Permagon Press.

Bilbro, T., Boni, M., Johnson, B., & Roe, S. (1979). *The relationship of parental acceptance-rejection, to the development of moral reasoning*. Unpublished manuscript. University of Connecticut, Center for the Study of Parental Acceptance-Rejection.

Bjorklund, D.F. (1988). Acquiring a mnemonic: Age and category knowledge effects. *Journal of Experimental Child Psychology,* Vol. 45, 71-87.

Boney-McCoy, S. & Finkelhor, D. (1995). The psychosocial sequelae of violent victimization in a national youth sample. *Journal of Consulting and Clinical Psychology,* 63, 726-736.

Bousfield, W.A. (1953). The occurrence of clustering in the recall of randomly arranged associates. *Journal of General Psychology.* 1953, 49, 229-240.

Bousha, D.M. & Twentyman, C.T. (1984). Mother-child international style in abuse, neglect and control groups: Naturalistic observations in the home. *Journal of Abnormal Psychology*, 93, 106-114.

Bower, G.H. (1981). Mood and memory. *American Psychologist*, 36, 129-148.

Bowlby, J. (1969). *Attachment and loss*: Vol.1. Attachment, New York: Basic books.

Bowlby, J. (1973). *Attachment and loss*: Vol. 2 Separation. New York: Basic Books.

Bowlby, J. (1980). *Attachment and loss*: Vol. 3, Loss, Sadness and Depression, New York: Basic Books.

Brassard, M.R., Hart, S.N., & Hardy, D. (1991). Psychological and emotional abuse of children. In R.T. Ammerman & M. Hersen (Eds), *Case Studies in Family Violence,* (pp. 255- 270). New York: Plenum.

Briere, J. & Runtz, M. (1988). Multivariate correlates of childhood psychological and physical maltreatment among university women. *Child Abuse & Neglect*, 12, 331-341.

Briere, J. & Runtz, M. (1990). Differential adult's symptomatology associated with three types of child abuse histories. *Child Abuse & Neglect*, 14, 357-364.

Briere, J. (1997). 'An integrated approach to treating adults abused as children with specific reference to self-reported recovered memories,' in D. Read and S. Lindsay (Eds.). *Recollections of Trauma: Scientific Research and Clinical Practices*, New York: Plenum. pp. 25-41.

Brinkerhops, M.B. & Lupri, E. (1988). Interpersonal violence. *Canadian Journal of Sociology*, 13, 407-434.

Broadbent, A. & Bentley, R. (1997). Child abuse and neglect Australia, 1995-96 (Child Welfare Series, No. 17; AIHW Cat No. CWSI). Canberra: Australian Institute of Health and Welfare.

Bronfenbrenner, U. (1977). Towards an experimental ecology of human development. *American Psychologist*. 32, 513-531.

Bronfenbrenner, U. (1979). The ecology of human development: experiments by nature and design. Cambridge, M.A.: Harvard University Press.

Brooks-Gunn, J. & Duncan, G. J. (1997). The effects of poverty on children. *The future of children*, 7 (2), 55-71.

Browne, A. & Finkelhor, D. (1986). Impact of child sexual abuse: A review of the research. *Psychological Bulletin*, 99, 66-77.

Burgess, R.L. (1985). Social incompetence as a precipitant to and consequences of child maltreatment. *Victomology: An International Journal,* 10, 72-86.

Calfee, R.C. (1975). *Human Experimental Psychology*. N.Y.: Holt.

Cappel, C. & Heiner, R.B. (1990). The intergenerational transmission of family aggression. *Journal of Family Violence*. 5, 135-152.

Carlson, B.E. (1986). Children's beliefs about punishment. *American Journal of Orthopsychiatry*, 56, 308-312.

Carlson, B.E. (1991). Emotionally disturbed children's beliefs and punishment. *Child Abuse & Neglect*, 15, 19-28.

Carver, C.S., Scheiver, M.F. & Weintraub, J.K. (1989). Assessing coping strategies. A Theoretically based approach, *Journal of Personality and Social Psychology*. 56, 267-283.

Cattell, R.B. (1950). Handbook for the Individual or Group Culture Fair (or Free) Intelligence Test, Scale 1. Institute for Personality and Ability Testing. Champaign, Illinois, U.S.A. by The Psycho Center T-22, New Delhi.

Cattell, R.B. (1972). Handbook for the Children's Personality Questionnaire. Institute for Personality and Ability Testing. Champaign, Illinois, U.S.A. by The Psycho Center T-22, New Delhi.

Cattell, R.B. (1972). Technical Supplement for the Culture Fair Intelligence Test, Scale 2&3. Institute for Personality and Ability Testing. Champaign, Illinois, U.S.A. by The Psycho Center T-22, New Delhi.

Cattell, R.B. (1973). Manual for Measuring Intelligence with the Culture Fair Intelligence Test, Scale 2&3. Institute for Personality and Ability Testing. Champaign, Illinois, U.S.A. by The Psycho Center T-22, New Delhi.

Cattell, R.B. (1973). Manual for the Jr.-Sr. High School Personality Questionnaire. Institute for Personality and Ability Testing. Champaign, Illinois, U.S.A. by The Psycho Center T-22, New Delhi.

Cattell, R.B., Saunders, D.R., & Stice, G.F. (1950). The 16 Personality Factor Questionnaires. Champaign, Il: Institute of Personality and Ability Testing.

Cerezo, M.A. (1997). Abusive Family Interaction: A review. *Aggression and Violent Behavior*, 2(3), 215-240.

Chan, J. (1981). Correlates of parent-child interaction and certain psychological variables among adolescents in Hong Kong. In J.L.M. Dawson et al. (Eds). *Perspectives in Asian Psychology*. Lisse: Swets & Zeitlinger.

Chan, Y.C. (1994). Parenting stress and social support of mothers who physically abuse their children in Hong Kong. *Child Abuse & Neglect*. 18, 261-269.

Chattoraj, R.N. & Saxena, R. (1990). Victimization of children: An urgent need for effective measures. *Social Change*. Vol. 20. No. 3, 17-23.

Christopoulos, C., Bonvillian, J. D. & Crittenden, P.M. (1988). Maternal language input and child maltreatment. *Infant Mental Health Journal*, 9, 272-286.

Cicchetti, D., & Olsen, K. (1990). The developmental psychopathology of child maltreatment. In M. Lewis & S.M. Miller (Eds), *Handbook of Developmental Psychopathology*, (pp. 261-279). New York: Plenum.

Claussen, A. H. & Crittenden, P. M. (1991). Physical and psychological maltreatment: Relations among the types of maltreatment. *Child Abuse & Neglect*, 15, 15-18.

Colvin, W.E. (1971). An evaluation of visual/verbal discriminative treatment upon low socio-economic status. *Dissertation Abstracts*, 32 (5A), 2298.

Connor, M. (1987). Treatment programs for abused children. NCPCA working paper no. 37, Chicago: National Committee for Prevention of Child Abuse.

Conte, J.R. & Schuerman, J.R. (1987). Factors, associated with an increased of child sexual abuse. *Child Abuse & Neglect*, 11, 201-211.

Coohey, C. (1996). Child maltreatment: Testing the social isolation hypothesis. *Child Abuse & Neglect*, 20, 241-254.

Costa, P.T., Jr. & McCrae, R.R. (1997). Stability and change in personality assessment: The Revised NEO Personality Inventory in the year 2000. *Journal of Personality Assessment*, 68, 86-94.

Coster, W.J., Gersten, M.S., Beeghly, M. & Cicchetti, D. (1989). Communicative functioning in maltreated toddlers. *Development and Psychopathology*, 25, 777-793.

Coulton, C. J., Korbin, J.E., Su, M, & Chow J. (1995). Community level factors and child maltreatment rates. *Child Development*, 66. 1262-1276.

Courtois, C.A., (1997). Informed clinical practice and the standard of care: proposed guidelines for treatment trauma; in D. Read and S. Lindsay (Eds.). *Recollections of Trauma: Scientific Research and Clinical Practice.* New York: Plenum. pp. 337-61.

Crittenden, P.M. (1988). Distorted patterns of relationship in maltreating families: The role of internal representation models. *Journal of Reproductive and Infant Psychology*, 6, 183-199.

Crittenden, P.M. (1992). Children's strategies for coping with adverse home environments: An interpretation using attachment theory. *Child Abuse & Neglect,* 16 (3), 329-339.

Crittenden, P.M. (1993). Characteristics of neglectful parents: An information processing approach. *Criminal Justice and Behavior*, 20, 27-48.

Crittenden, P.M. (1998). Dangerous behavior and dangerous contexts: A 35-year perspective on research on the development effect of child physical abuse. In P.K. Trickett, & C.J. Schelienback (Eds.). *Violence against children in the family and the community* (pp. 11-38). Washington, D.C., American Psychological Association.

Crittenden, P.M., Ainsworth, M.D.S. (1989). Child maltreatment and attachment theory. In D. Cicchetti & V. Carlson (Eds.) *Child maltreatment: Theory and research on the causes and consequences of child abuse and neglect* (pp. 423-463). New York: Cambridge University Press.

Crittenden, P.M., Claussen, A. H. & Sugarman, D.B. (1994). Physical and psychological maltreatment in middle childhood and adolescence. *Development and Psychopathology*, 6, 145-164.

Cummings, E.M., Hennessy, K.D., Radideau, G.J. & Cicchetti, D. (1994). Responses of physically abused boys to interadult anger involving their mothers. *Development and Psychopathology*, 6 (1), 31-41.

D'Amato, M.R. (1970). *Experimental Psychology*, N.Y. McGraw Hill.

Dash, M. (1991). Analysis of cognitive and speech-related processes in relation to reading efficiency and I.Q. M.Phill, Psy, Utkal University.

De Bellis, M. D., Chrousos, G. P., Dorn, L.D., Burke, L., Helmers, K., Kling, M.A., Trickett, P.K. & Putnam, F.W. (1994). Hypothalamic-Pituitaryadrenal axis dysregulation in sexually abused girls. *Journal of Clinical Endocrinology and Metabolism*, 78, 249-255.

De Bellis, M.D., Burke, L., Trickett, P.K., & Putnam, F.W. (1996). Antinuclear antibodies and thyroid function in sexually abused girls. *Journal of Traumatic Stress*, 9, 369-378.

De Board, L.W. (1969). The achievement syndrome among Negro and White culturally disadvantaged boys. *Unpub. Ph.D.* Thesis, Vanderbidt University.

De Francis, V. (1969). *A Research Protecting the Child Victim of Sexual Crimes Committed by Adults*. Project conducted under child welfare research grant R-222 for the U.S. Children's Bureau. Denver: American Human Association.

De Paul, J., & Arruabarrena, M. I. (1995). Behavior problem in school aged physically abused and neglected children in Spain. *Child Abuse & Neglect*, 19, 409-418.

Dei, S.L. (1991). Interrelationship between non-verbal measures of cognition. M.Phil. Psy, Utkal University.

Department of Health and Human Services. (1981). *Study Findings: National Study of the Incidence and Severity of Child Abuse and Neglect* (DHHS Publication no. OHDS, 81-30325). Washington, DC: Government Printing Office.

Department of Health and Human Services. (1988). *Study Findings: National Study of the Incidence and Severity of Child Abuse and Neglect* (DHHS Publication no. ADM 20-01099). Washington, DC: Government Printing Office.

Department of Health and Human Services. (1993). *A Report on the Maltreatment of Children with Disabilities* (DHHS Publication no. 105-89-1630). Washington, DC: Government Printing Office.

Derogowski, J.B. (1980). Illusion patterns and pictures: A cross-cultural perspective. New York: Academic.

Dixit, S. (1988). *Information's Processing: An Analysis of the acquisition of Learning*. Ph.D. Psy. The Maharaja Sayajirao University of Boroda.

Dodge, K.A., Bates. J. E. & Pettiet, G.S. (1990). Mechanisms in the cycle of violence. *Science*, 250, 1678-1682.

Dodge, K.A., Pettit, G.S. & Bates, J.E. (1997). How the experience of early physical abuse leads children to become chronically aggressive. In *developmental perspectives on trauma: Theory, research and intervention* (Vol.8, pp. 263-288). Rochester, N. Y.: University of Rochester Press.

Dollard, J., & Miller, N.E. (1950). *Personality and Psychotherapy*. New York: McGraw-Hill.

Dollard, J., Doob, L.W., Miller, N.E., Mowrer, O.H., & Sears, R.R. (1939). *Frustration and Aggression*. New Haven, C.T. Yale University Press.

Donga, K.M. (1989). A study of memory and affecting variables of students studying in standards VIII to XII. Ph.D. Edu. Saurashtra.

Downs, W.R. & Miller. B.A. (1998). Relationship between experiences of parental violence during childhood and women's psychiatric symptomatology. *Journal of Interpersonal Violence*. 13 (4), 438-455.

Drake, B. & Pandey, S. (1996). Understanding the relationship between neighborhood poverty and specific types of child maltreatment. *Child Abuse and Neglect*, 20, 1003-1018.

Dyk, R., B. & Witkin, H.A. (1965). Family experiences related to the development of differentiation in children. *Child Development*, 36, 21-55.

Eckenrode, J., Laird, M. & Doris, J. (1993). School performance and disciplinary problems among abused and neglected children. *Developmental Psychology*, 29, 53-63.

Edleson, J.L. (1999). The overlap between child maltreatment and women bettering. *Violence Against Women*, 5, 134-154.

Edleson, J.L., Eisikovits, Z.C., Guttmann, E., & Sela-Amit, M. (1991). Cognitive and interpersonal factors in woman abuse. *Journal of Family Violence*, 6, 167-182.

Egan, K.C. (1983). Stress management and child management with abusive parents. *Journal of Clinical Child Psychology*, 12, 292-299.

Egeland, B. & Erickson, M. (1987). Psvchologically unavailable care giving. In M.R. Brassard, R. Germain & S.N. Hart (Eds),

Psychological Maltreatment of Children and Youth (pp. 110-120). New York: Pergamon.

Egeland, B. & Sroufe, L.A. (1981). Developmental sequeale of maltreatment infancy. In B. Rizely & D. Cicchetti (Eds), *New Directions for Child Development: Developmental Perspectives in Child Maltreatment*, 11, (pp. 77- 92). San Francisco: Jossey Bass.

Egeland, B. (1991). From data to definition. *Development and Psychopathology*, 3, 37-43.

Egeland, B., Sroufe, L.A. & Erickson, M.F. (1983). The developmental consequences of different patterns of maltreatment. *Child Abuse & Neglect*, 7, 459-469.

Einbender, A. J., & Friedrich, W.N. (1989). Psychological functioning and behavior of sexually abused girls. *Journal of Consulting & Clinical Psychology*, 57,155-157.

Erickson, M.F., Egeland, B. & Pianta, R. (1989). The effects of maltreatment on the development of young children. In D. Cicchetti & V. Carlson, (Eds.). *Child Maltreatment: Theory and Research on the Causes and Consequences of Child Abuse and Neglect,* (pp. 647-684). New York: Cambridge University Press.

Ethier, L.S., Lacharite, C. & Couture, G. (1995). Childhood adversity, parental stress and depression of negligent mothers. *Child Abuse & Neglect*, 19, 619-632.

Everson, M.D., Hunter, W.M., Runyan, D.K., Edelson, G.A. & Coulter, M. L. (1989). Maternal support following disclosure of incest. *American Journal of Orthopsychiatry*, 59, 197-207.

Eysenck, S.B. & Eysenck, H.J. (1963). The validity of questionnaire and rating assessments of extraversion and neuroticism, and their factorial stability. *British Journal of Psychology*, 54, 51-62.

Faller, (1981). Causes of child abuse and neglect. In Faller K. (Ed). *Social work with abused and neglected children: A Manual of Interdisciplinary Practice*. New York: Macmillan.

Famularo, R., Fenton, T. & Kinscherff, R.T. (1992). Medical and developmental histories of maltreated children. *Clinical Pediatrics,* 31, 536-541.

Famularo, R., Kinscherff, R. & Fenton, T. (1991). Post-traumatic stress disorder among children clinically diagnosed as borderline personality disorder. *Journal of Nervous and Mental Diseases,* 179, 428-431.

Fantuzzo, J.W. (1990). Behavioral treatment of the victims of child abuse and neglect. *Behavior Modification*, 14, 316-339.

Fantuzzo, J.W., Depalola, L.M., Lambert, L., Martino, T., Anderson, G., & Sutton, S. (1991). Effects of interparental violence on the psychological adjustment and competencies of young children. *Journal of Consulting and Clinical Psychology,* 59, 258-265.

Farrington, K.M. (1980). Stress and Family Violence. In M.A. Straus & G.T. Hotaling (Eds.). *The social causes of husband-to-wife violence* (pp. 94-114). Minneapolis: University of Minnesota Press.

Feather, N.T. (1961). The relationship of persistence at a task to expectation of success and achievement related motives. *Journal of Abnormal and Social Psychology*, 63, 552-561.

Feldman, R.S., Salzinger, S., Rosario, M., Alvarado, L. & Hammer, M. (1995). Parent, teacher and peer ratings of physically abused and normal treated children's behavior. *Journal of Abnormal Child Psychology,* 23, 317-334.

Festinger, L. (1942). A theoretical interpretation of shift in level of aspiration, *Psychological Review*, 49, 235-250.

Finkelhor, D. & Browne : A. (1988). Assessing the long-term impact of child sexual abuse. A review and conceptualization, in L. Walker (Ed.). *Handbook on Sexual Abuse of Children* (pp. 55-71). New York: Springer.

Finkelhor, D. & Korbin, J. (1988). Child abuse as an international issue. *Child Abuse and Neglect.* 12 (1), 3- 23.

Finkelhor, D. (1993). Epidemiological factors in the clinical identification of child sexual abuse. *Child Abuse & Neglect,* 17, 67-70.

Finkelhor, D. (1994-a). Current information on the scope and nature of child sexual abuse. *Future of Children*, 4 (2), 31-53.

Finkelhor, D. (1994-b). The International epidemiology of child sexual abuse. *Child Abuse & Neglect*, 18, 409-417.

Finkelhor, D. (1995). The victimization of children: A developmental perspective. *American Journal of Orthopsychiatry*, 65, 177-193.

Finkelhor, D. (1996). Introduction. In J. Briere, L. Berliner, J.A. Bulkley, C. Jenny & T. Reid (Eds). *The APSAC Handbook on Child Maltreatment,* (pp. ix-xiii). Thousand Oaks, CA: Sage.

Fisher, A.J., Kramer, R.A., Hoven, C.W., Greenwald, S., Alegria, M., Bird, H.R., Canino, G., Connell, R., & Moore R.E. (1997). Psychosocial characteristics of physically abused children and adolescents. *Journal of the American Academy of Child and Adolescent Psychiatry,* 36, 123-131.

Ford, J.D., Racusin, R., Daviss, W.B., Ellis, C., Thomas, J., Rogers, K., Reiser, J., Schiffman, J., & Sengupta, A. (1999). Trauma exposure among children with attention deficit hyperactivity disorder and oppositional defiant disorder. *Journal of Consulting & Clinical Psychology*, 67, 786-789.

Frank, J.D. (1935). Individual differences in certain aspects of the level of aspiration. *American Journal of Psychology*, 47, 119-128.

Frankel, K.A. & Bates, J.E. (1990). Mother-toddler problem solving: Antecedents in attachment home behavior and temperament. *Child Development*, 61, 810-819.

Friedrich, W.N. (1990). *Psychotherapy of Sexually Abused Children and their Families*. New York: Norton.

Friedrich, W. N., Enbender, A.J., & Luecke, W.J. (1983). Cognitive and behavioral characteristics of physically abused children. *Journal of Consulting and Clinical Psychology*, 51, 313-314.

Friedrich, W.N. (1993). Sexual victimization and sexual behavior in children: A review of recent literature. *Child Abuse & Neglect,* 17, 59-66.

Friedrich, W.N., Dittner, C.A., Action, R., Berliner, L., Butler, J., Damon, L., Davies, W.H., Gray, A., & Wright, J. (2000). Child sexual behavior inventory: Normative, psychiatric and sexual abusive families: A behavioral comparison. *Journal of Interpersonal Violence*, 2, 391-402.

Friedrich, W.N., Jaworski, T. M., Huxsahl, J.E., & Bengston, B.S. (1997). Dissociative and sexual behaviors in children and adolescents with sexual abuse and psychiatric histories. *Journal of Interpersonal Violence*, 12. 155-171.

Friedrich, W.N., Urquiza, A.J., & Beilke, R. (1986). Behavioral problems in sexually abused young children. *Journal of Pediatric Psychology,* 11, 47-57.

Gaines, R., Sandgrund, A., Green, A.H., & Power, E. (1978). Etiological factors in child maltreatment: A Multivariate study of abusing, neglecting and normal mothers. *Journal of Abnormal Psychology,* 87, 531-540.

Gallup, Organization. (1995). *Discipline Children in America: A Gallup Poll Report.* Princeton, N.J. Author.

Garbarino, J. (1976). A preliminary study of some ecological correlates of child abuse: The impact of socio-economic-stress on mothers. *Child Development*, 47, 178-185.

Garbarino, J. (1997). The role of economic deprivation in the social context of child maltreatment. In M.E. Helfer, R.S. Kempe & R.D. Krugman (Eds). *The Battered Child,* (5th ed, pp. 49-60). Chicago: University of Chicago Press.

Garbarino, J. & Kostelny, K. (1992). Child maltreatment as a community problem. *Child Abuse and Neglect*, 16, 455-464.

Garg, K.P. (1988). A factorial study of reasoning abilities at the age of 14th+. Ph.D., Edu. Jamia Millia Islamia.

Gelles, R.J. (1983). An exchange/social control theory. In D. Finkelhor, R. J. Gelles, G. T. Hotalling, & M.A. Straus (Eds). *The Dark Side of Families: Current Family Violence Research,* (pp. 151-165). Beverly Hills CA : Sage.

Gelles, R.J. (1989). Child abuse and violence in single-parent families: Parent absence and economic deprivation. *American Journal of Orthopsychiatry*, 59 (4), 492-501.

Gelles, R.J., & Straus, M.A. (1988). Intimate violence: The definitive study of the causes and consequences of abuse in the American family. New York: Simon & Schuster.

Gibbs, J. (1975). *Crime, Punishment, and Deterrence*. New York: Elsevier.

Gil, D.G. (1970). *Violence against children: Physical abuse in the United States*. Cambridge, M.A. Harvard University Press.

Gil, D.G. (1971). Violence against children. *Journal of Marriage and Family*, 33, 637-648.

Girardau, F.L. (1959). The formation of discrimination learning sets in Mongoloid and normal children. *Journal of Comparative & Physiological Psychology*, 52, 566-570.

Glad Stone, G., Parker, G., Wilhelm, K., Mitchell, P. & Austin, M.P. (1999). Characteristics of depressed patients who report child sexual abuse. *American Journal of Psychiatry*, 156-812.

Gomes-Schwartz, B., Horotwitz, J.M. & Cardarelli, A. P. (1990). *Child Sexual Abuse: The Initial Effects*. Newbury Park, C.A.: Sage.

Green, D.R. (1963). Volunteering and the recall of interrupted task. *Journal of Abnormal and Social Psychology*, 66, 397-401.

Gross, A.B. & Keller, H.R. (1992). Long-term consequences of childhood physical and psychological maltreatment. *Aggressive Behavior*, 18, 171-185.

Gupta, G. C. (1980). Cognitive Processes. In U. Pareek (Ed). *A survey of research in psychology*, 1971-76, Vol-1, Bombay: Popular.

Gupta, S. (1991). A study of deprivation in relation to certain cognitive and non-cognitive variables among adolescents. Ph.D., Edu. University of Jammu.

Harlow, H.F. (1949). The formation of learning sets. *Psychological Review*, 56, 51-65.

Harlow, H.F., (1960). Learning set and error factor theory. In Koch, S. (Ed.). *Psychology: A Study of a Science*. Vol.2. N.Y.: McGraw-Hill, pp. 492-537.

Hart, S.N. & Brassard, M.R. (1991). *Psychological Maltreatment: Progress Achieved Development and Psychopathology*, 3, 61-70.

Hart, S.N. & Brassard, M.R. (1993). *Psychological Maltreatment*. Violence Update, 3(7). N. 4, 6-7, 11.

Haskett, M. (1990). Social problem solving skill of young physically abused children. *Child Psychiatry and Human Development*, 21, 109-118.

Heaven, P.C. (1996). *Personality and Self-reported Delinquency: Analysis of the " Big Five" Personality Dimensions, Personality and Individual Differences*, 20, 47-54.

Herrenkohl, E.C., Herrenkohl, R.C., Rupert, L.J., Egolf, B.P. & Lutz, J.G. (1995). Risk factor for behavioral dysfunction: The relative impact of maltreatment, SES physical health problems, cognitive ability and quality of parent child interaction. *Child Abuse & Neglect*, 19, 191-203.

Herrenkohl, R.C., Herrenkohl, E.C. & Egolf, B.P. (1983). Circumstances surrounding the occurrence of child maltreatment. *Journal of Consulting and Clinical Psychology*, 51, 424-431.

Herrenkohl, R.C., Herrenkhol, E.C., Egolf, B.P. & Wu. P. (1991). *The Development Consequences of Child Abuse*. The light longitudinal study (Grant MH 4110) Washington, D.C.: National Center on Child Abuse and Neglect.

Hilton, Z. N. (1992). Battered women's concerns about their children-witnessing wife assault, *Journal of Interpersonal Violence*, 7, 77-86.

Hirschi, T. (1969). *Causes of Delinquency*. Berkeley: University of California Press.

Hoffman, M.L. (1970). Moral Development, In P. H. Mussen (Ed.), *Carmichael's manual of child psychology* (Vol-2). New York: John Wiley.

Hoffman-Poltkin, D., & Twentyman, C.T. (1984). A multi-model assessment of behavioral and cognitive deficits in abused and neglected preschoolers. *Child Development*, 55, 794-802.

Hoglund, C.L. & Nicholas, K.B. (1995). Shame guilt and anger in college students exposed to abusive family environments. *Journal of Family Violence*, 10(2), 141-157.

Hollowitz, D., & Stulberg, B. (1959). The vicious cycle of parent-child relationship breakdown. *Social Case Work*. 40, 268-275.

Holmes, T.H. & Rahe, R.H. (1967). The social readjustment rating scale, *Journal of Psychosomatic Research*, 11, 213-218.

Hoppe, F. (1930). Untersuchungenzur handlungs und affect-psychologie. Ix. Erfolg und misserfolg. *Psychologische Forschung* 14, 1-63.

Hotaling, G.T., Straus, M.A. & Lincolin, A. J. (1990). Intra Family Violence and Crime and Violence Outside the Family. In M.A. Straus & R.J. Gelles (Eds). *Physical violence in American families. Risk factor and adaptation to violence in 8,145 families* (pp. 413-470). New Brunswick . N.J. Transaction Books.

Howes, C., & Espinosa, M.P. (1985). The consequences of child abuse for the formation of relationship with peers. *Child Abuse & Neglect*, 9, 397-404.

Hughes, H.M. & Graham-Bermann, S.A. (1998). Children of battered women: Impact of emotional abuse on adjustment and development. *Journal of Emotional Abuse,* 1(2), 23-50.

Hunter, J., Goodwin, D.W. & Wilson, R. J. (1992). Attributions of blame in child sexual abuse victims: An analysis of age and gender influences. *Journal of Child Sexual Abuse*, 1, 75-90.

Hunter, W.M., Coulter, M.L., Runyan, D.K. & Everson, M.D. (1990). Determinants of placement for sexually abused children. *Child Abuse & Neglect*, 14, 407-418.

Hurlock, E.B. (1984). *Child Development*. McGraw-Hill series in Psychology.

Institute of Medicine. (1989). *Research on children and adolescents with mental, behavioral and developmental disorders.* Washington, D.C.: National Academy Press.

Irvine, S.H. & Berry, J.W. (1988). *Human Abilities in Cultural Context*, New York: Oxford University Press.

Jensen, A.R. (1965). *Individual Differences in Learning: Interference Factor*. Co-operative Research Project No.1867, United States Office of Education.

Joshi, M.C., & Vyas, R. (1997). *Manual of Directions for Family*

Environment Scale. Rupa Psychological Center, Bhelupur, Varanasi.

Kaplan, S., Pelcovitz, D., Salzinger, S., Mandel, F. S. & Weiner, M. (1998). Adolescent physical abuse. Risk for adolescents psychiatric disorders. *American Journal of Psychiatry*, 155(7), 954-959.

Kaufman, J. & Cicchetti, D. (1989). Effect of maltreatment on school age children's socio-emotional development: Assessment in day camp setting. *Developmental Psychology*, 25, 516-524.

Kaufman, J. & Zigler, E. (1987). Do abused children become abusive parents? *American Journal of Orthopsychiatry*, 57, 186-192.

Kempe, C., & Helfer, R. (1972). *Helping the battered child and his family.* Philadelphia: J. B. Lippinott.

Kempe, C.H., Silverman, F.N., Steel, B.F., Drogemueller, W. & Silver, H.K. (1962). The bartered child syndrome. *Journal of the American Medical Association*, 17, 17-24.

Kendall, P. C., Stark, K. D. & Adam, T. (1990). Cognitive deficit or cognitive distortion in childhood depression. *Journal ofAbnormal Child Psychology*, 18, 255-270.

Kendall, P.C. (1993). Cognitive-behavioral therapies with youth: Guiding theory, current status, and emerging developments. *Journal of Consulting and Clinical Psychology*, 61, 235-247.

Kendall-Tackett, K.A., Williams, L.M. & Finkelhor, D. (1993). Impact of sexual abuse on children: A review and synthesis of recent empirical studies. *Psychological Bulletin*, 113, 164-180.

Kendall-Tackett, K.A., Williams, L.M., & Finkelhor, D. (1993). Impact of sexual abuse on children: A review and synthesis of recent empirical studies. *Psychological Bulletin*, 113, 164-180.

Kennedy, J.M. (1974). *A Psychology of Picture Perception.* San Fransicso: Jossey-Bass.

Kent, A. & Waller, G. (1998). The impact of childhood emotional abuse: An extension of the child abuse and trauma scale. *Child Abuse & Neglect.*

Kilpatric, D.G. & Saunders, B. E. (1999). Prevalence and consequence of child victimization: Result from the national survey of adolescents (No. 93-IJ- CX-0023). Charleston: National Crime Victims Research & Treatment Center, Department of Psychiatric & Behavioral Science, Medical University of South Carolina.

Kinard, E.M. (1982). Experiencing child abuse. Effects on emotional adjustment. *American Journal of Orthopsychiatry*, 50, 686-696.

Klimes-Dougan, B. & Kistner, J. (1990). Physically abused preschoolers' response to peers' distress. *Developmental Psychology*, 26, 599-602.

Knutson, J.F. (1978). Child abuse as an area of aggression research. *Journal of Pediatric Psychology*, 3, 20-27.

Kolko, D.J. (1992). Characteristics of child victims of physical violence: Research finding and clinical implications. *Journal of Interpersonal Violence*, 7, 244-276.

Kolko, D. J., Moser, J. & Wieldy, S.R. (1988). Behavioral/emotional indicators of sexual abuse in psychiatric inpatients: A controlled comparison with physical abuse. *Child Abuse & Neglect*, 12, 529-541.

Korbin, J. (1987). Child maltreatment in cross-cultural perspective: Vulnerable children and circumstances. In R.J. Gelles & J.B. Lancaster (Eds.), *Child Abuse & Neglect: Biosocial Dimensions* (pp. 31-55). New York: Aldine de Gruyter.

Korbin, J. (1994). Socio-cultural Factors in Child Maltreatment. In G.B., Melton & F. D. Barry (Eds.), *Protecting children from abuse and neglect. Foundations for a new national strategy*, (pp. 182-223). New York: Guildford.

Korbin, Jill E. (1981). *Child Abuse and Neglect*, University of California, p. vil.

Kukla, A. (1972). Attributional determinants of achievement related behavior. *Journal of Personality and Social Psychology*, 21, 166-174.

Kumari, D. (1986). Intellectual commitment and educational interest in relation to certain cognitive and non-cognitive variables. Ph.D. Edu. University of Jammu.

Kurtz, P.D., Gaudin, J. M., Wodarski, J.S. & Howing, P.T. (1993). Maltreatment and the school aged child: School performance consequences. *Child Abuse & Neglect*, 17, 518-589.

Kurtz, P.D., Gaudin, J.M., Howing, P.T., & Wodarski, J.S. (1993). The consequences of physical abuse and neglect on the school age child: Mediating factors: *Children and Youth Services Review*, 15, 85-104.

Larrance, D. T., & Twentyman, C.T. (1983). Maternal attribution and child abuse. *Journal of Abnormal Psychology*, 92, 449-457.

Latane, B. & Darley, J.M. (1970). *The Unresponsive Bystander: Why Doesn't be Help*? New York: Appleton-Century Crofts. Reviews the classic studies of bystander intervention conducted by these authors.

Lazarus, R. & Folkman, S. (1984). *Stress, Appraisal and Coping*. New York : Springer Publishing.

Lazarus, R.S. (1981). The Stress and Coping Paradigm. In C. Eisdoefer, D. Cohen, A. Kleinamn, & P. Maximum (Eds). *Models for Clinical Psychopathology* (pp. 177-214). New York: Spectrum.

Lehman, H.C. (1968). The creative production rates of present versus past generation of scientists. In B.L. Naugarten (ed.), *Middle Age and Aging: A Reader in Social Psychology*. Chicago: University of Chicago Press, pp. 99-105.

Lembright, M.L. & Yamamoto, K. (1965). Subcultures and creative thinking: An exploratory comparison between Amish and Urban American school children. *Merrill-Palmer Quarterly*, 1965, 11, 49-64.

Light, R. (1973). Abused and neglected children in America: A study of alternative policies. *Harvard Educational Review*, 43, 556-598.

Limber, S.P. & Nation, M.A. (1998). Violence within the neighborhood and community. In P.K. Trickett, & C.J. Schellenbach, (Eds.), *Violence Against Children in the Family and the Community* (pp. 171-193). Washington. DC: American Psychological Association.

Lindberg, F.H. & Distad, L.J. (1985). Post-traumatic stress disorder in women who experienced childhood incest, *Child Abuse and Neglect*, 9: 329-34.

Lindsay, D.S. & Briere, J. (1997). The controversy regarding recovered memories of childhood sexual abuse: Pitfalls, bridges and future directions. *Journal of Interpersonal Violence*, 12, 631-647.

Lutzker, J.R., Campbell, R.V. & Watson-Perozel, M. (1984). Using the case study method to treat several problems in a family indicated for child neglect. *Education and Treatment for Children's*, 141-155.

Maccoby, E.E. & Martin, J.A. (1983). Socialization in the context of the family: Parent–child interaction. In P.H. Mussen (Ed), *Handbook of Child Psychology*. (4th Ed. pp. 1-101). New York: John Wiley.

Mahone, C.H. (1960). Fear of failure and unrealistic vocational aspiration. *Journal of Abnormal and Social Psychology*. 60, 253-261.

Main, M. & George, C. (1985). Responses of abused and disadvantaged toddlers to distress in age mates: A study in the day care setting. *Developmental Psychology*, 21(3), 407-412.

Main, M. & Goldwyn, R. (1984). Predicting rejection of her infant from mother representation of her own experiences: Implication for the abuse-abusing intergenerational cycle. *Child Abuse & Neglect.* 8, 203-217.

Malamuth, N. M., Sockloskie, R.J., Koss, M.P., & Tanaka, J.S. (1991). Characteristics of aggressors against women: Testing a model using a national sample of college students. *Journal of Consulting and Clinical Psychology*, 59, 670-681.

Manley, R. (1977). Parental warmth and hostility as related to sex differences in children's achievement orientation. *Psychology of Women Quarterly*, 1, 129-246.

Manly, J.T., Cicchetti, D. & Barnett, D. (1994). The impact of subtype, frequency, chronicity, and severity of child maltreatment on social competence and behavior problems. *Development and Psychopathology*.

Markus, H.R. & Kitayama, S. (1998). The Cultural Psychology of Personality. *Journal of Cross-Cultural Psychology*, 29(1), and 63-87.

Masten, A. S. & Garmezy, N. (1985). Risk, vulnerability, and protective factors in developmental psychopathology. In B.B. Lahey & A.E. Kazadin (Eds.), *Advances in Clinical Child Psychology*. (Vol. 8, pp. 1-52). New York: Plenum.

Mayer, H.H., Walker, W.B. & Litwin, G.H. (1961). Motive patterns and risk preference associated with entrepreneurship. *Journal of Abnormal and Social Psychology,* 63, 570-574.

McClelland, D.C. (1961). *The Achieving Society*, N.J., D. Van Nostrand.

McClelland, D.C. & Winter D.G. (1969). *Motivating Economics Achievement.* N.Y. the free Press.

McClelland, D.C. Atkinson, J.W., Clark, R.W.S. & Lowell. E.C. (1953). *The Achievement Motive*. N.Y.: Appleton-Century-Crofts.

McCrae, R.R., & Costa, P.T. Jr. (1990). *Personality in Adulthood.* New York: Guilford Press.

McFayden, R.G. & Kitson, W.J.H. (1996). Language comprehension and expression among adolescent who have experienced childhood physical abuse. *Journal of Child Psychology and Psychiatry and Allied Disciplines*, 37, 551-562.

McLoyd, V.C. (1998). Socio economic disadvantage and child development. *American Psychologist*, 53, 185-204.

Mehdi, B. (1973). *Verbal and Non-Verbal Test of Creative Thinking*. National Psychological Corporation. Raja Mandi, Agra-2

Mehta, M. (1979). Kidnapping: A Social Evil. In R. Helfer & R. Helfer & C. Kempe (Eds.). *Child Abuse and Neglect,* Vol. 3, 678-682.

Mehta, P. (1978). *A Brief Manual for Achievement Motivation Test.* Manasayan, 32, Netaji Subhash Marg, New Delhi.

Mickelson, N.I. & Galloway, C.G. (1973). Verbal concepts of Indian and non-Indian school beginners. *The Journal of Educational Research*, 67-72.

Milner, J.S., Halsey, L.B. & Fultz, J. (1995). Empathic responsiveness and affective reactivity to infant in high and low risk for physical child abuse mothers. *Child Abuse & Neglect,* 19, 776-780.

Milner, J.S. & Wimberly, R.C. (1980). Prediction and explanation of child abuse. *Journal of Clinical Psychology*, 36, 875-884.

Milner, J.S. (1993). Social information processing and physical child abuse. *Clinical Psychology Review*, 13, 275-294.

Milner, J.S. (1998). Individual and family characteristics associated with interfamilial child physical and sexual abuse. In P. K. Trickett & C.J. Schellenbach (Eds). *Violence Against Children in the Family and Community* (pp.141-170). Washington, D.C., American Psychological Association.

Milner, J.S. & Chilamkurti, C. (1991). Physical child abuse perpetrator characteristic: A review of the literature. *Journal of Interpersonal Violence*, 6, 336-344.

Milner, J.S., (1995). Physical child abuse assessment: Perceptual evaluation. In J.C. Campbell (Ed.). *Assessing for Dangerousness: Violence by Sexual Offenders, Batterers, and Child Abusers* (pp. 41-67). Thousand Oaks. CA: Sage.

Mischel, W. (1961). Delay of gratification need for achievement and acquiescence in another culture. *Journal of Abnormal and Social Psychology*, 62, 543-552.

Mishra, R.C. (1992). Fifth survey of educational research volume: *National Council of Educational Research and Training*.

Mishra, R.C. (1996). Caste, education and urban contact as factors in pictorial perception and interpretation. *Indian Educational Review,* 31, 17-30.

Mishra, R.C. (1996). Perceptual differentiation in relation to children's daily life activities. *Social Science International*, 12, 1-11.

Mishra, R.C. (1999). Research on education in India. In S. Perez & P.R. Dasen (Eds), *Open File: Educational Research, Special Issue of Propects, Quarterly Review of Comparative Education*, XXIX, 3 (pp. 335-348). Geneva: TBE/UNESCO.

Mishra, R.C. (2001). Cognition Across Cultures. In D. Matsumoto (Ed.). *The Handbook of Culture and Psychology* (pp. 119-135). New York, Oxford University Press.

Misra, G. & Tripathi, L.B. (1980). *Psychological Consequences of Prolonged Deprivation*. Agra: National Psychological Corporation.

Mohsin, S.M. (1979). *Mohsin Parent Child Inventory*. Rupa Psychological Center, Bhelupur, Varanasi.

Morris, J.L. (1966). Propensity for risk taking as a determinant of vocational choice: An extension of the theory of achievement motivation. *Journal of Personality and Social Psychology*, 3, 328-335.

Mukherjee, B.N. & Sinha, R. (1970). Achievement value and self-ideal discrepancies in college students. *Personality: An International Journal*, 1, 275-301.

Muthayya, B.C. (1971). Farmer's levels of aspiration. Hyderabad. NICD.

National Academy of Sciences (1993). *Understanding Child Abuse and Neglect*. Washington, D.C: National Academy Press.

National Advisory Mental Health Council. (1995). Basic behavioral science research for mental health. *American Psychologist*, 50, 838-845.

National Research Council (1993). *Understanding Child Abuse and Neglect*. Washington, D.C.: National Academy Press.

Neisser, U. (1967). *Cognitive Psychology*, N.Y.: Appleton Century Crofts.

Newberger, C.M. & Cook, S. J. (1983). Parental awareness and child abuse: A cognitive development analysis of urban & rural samples, *American Journal of Orthopsychiatry*, 53, 512-524.

Niraula, S. & Mishra R.C. (2001). Spatial orientation of the newar children in Nepal. *Social Science International*, 17, 36-48.

Nygard, R. (1969). Motive to approach success and motive to avoid failure in boys from different social groups. *Scandinavian Journal of Educational Research*, 4, 222-232.

O'Connor, S. Vietze, P., Sherrod, K., Sandler, H. & Altemeier, W. (1990). Reduced incidence of parenting inadequacy following rooming in Pediatrics, 66, 176-182.

O'Hagan, K. (1995). Emotional and psychological abuse: Problems of definition. *Child Abuse & Neglect*, 19, 449-461.

Okun, A., Parker, J.G., & Levendosky, A.A. (1994). Distinct and interactive contribution of physical abuse, socio economics disadvantage, and negative life events to children social, cognitive and affective adjustment. *Development and Psychopathology*, 6(1), 77-98.

Osborne, J. (1997). Race and academic disidentification. *Journal of Educational Psychology*, 89, 728-735.

Pandey, S. & Srivastava, S. (2000). Coping with work stress in career oriented females. Published paper in *Journal of Community Guidance and Research.* Vol. 17, No.3, pp. 313-323.

Pandey, S. & Srivastava, S. (2000). Coping with work stress: The role of job category, family type and job tenure published paper in *Journal of Research and Application in Clinical Psychology*, Vol. Iii (i & ii), 1-10.

Pandey, S. & Srivastava, S. (2002). Coping with work stress: The role of job category, family type and job tenure. *Paper presented at National Seminar on at Culture and Life Style.*

Pandey, S. & Srivastava, S. (2004). Role of coping in work stress-health relationship: A study on career women. Akbar Husain & Tabassum Rashid (Eds). *Stress Behavior: The Coping Strategies,* Vol.2.

Pandey, S. (2002). *Helping Behavior Questionnaire.* First annual report submitted to U.G.C., unpublished research project. Department of Psychology, D.S.A., D.D.U. Gorakhpur University, Gorakhpur.

Pandey, S. (2002). *Personal Data Sheet (P.D.S.).* First annual report submitted to U.G.C., unpublished research project. Department of Psychology, D.S.A., D.D.U. Gorakhpur University, Gorakhpur.

Pandey, S. (2002). *Aspiration Scale.* First annual report submitted to U.G.C., unpublished research project. Department of Psychology, D.S.A., D.D.U. Gorakhpur University, Gorakhpur.

Pandey, S. (2002). *Child Abuse and Neglect: Interview Schedule for Parents.* First annual report submitted to UGC unpublished, research project. Department of Psychology, D.S.A., DDU Gorakhpur University, Gorakhpur.

Pandey, S. (2002). *Child Abuse Checklist.* First annual report submitted to U.G.C., unpublished research project. Department of Psychology, D.S.A., D.D.U. Gorakhpur University, Gorakhpur.

Pandey, S. (2002). *Child Neglect Scale.* First annual report submitted to U.G.C. unpublished, research project. Department of Psychology, D.S.A., DDU Gorakhpur University, Gorakhpur.

Pandey, S. (2002). *Memory Span Test.* First annual report submitted to U.G.C. unpublished, research project. Department of Psychology, D.S.A., DDU Gorakhpur University, Gorakhpur.

Pandey, S. (2002). *Verbal Learning Task.* First annual report submitted to U.G.C. unpublished, research project. Department of Psychology, D.S.A., DDU Gorakhpur University, Gorakhpur.

Pandey, S. (2004). Impact of Environmental Setting and Child Abuse on Creative Potentials in Children. *Paper Presented in National Seminar on Psychological and Environmental Change.* Department of Psychology, D.S.A., DDU Gorakhpur University, Gorakhpur.

Pandey, S. (2004). *Preventing Child Maltreatment: Ecological Perspective.* Unpublished research project report, under D.S.A. programme. Department of Psychology (D.S.A.), D.D.U. Gorakhpur University, Gorakhpur.

Pandey, S. (2005). "Social Dynamics of Child Abuse and its Psychological Consequences in Eastern Districts of U.P." Unpublished U.G.C. Research Project. Department of Psychology (D.S.A.), D.D.U. Gorakhpur University, Gorakhpur.

Pandey, S. (2005b). Child abuse an impediment to the development of Creative potential in children. *Psychological Studies,* 15, 234-242.

Pati, G.C. (1976). *Aggression Questionnaire.* Rupa Psychological Center. Varanasi.

Paunonen, S.V., Jackson, D., Trzebinski, J. & Forterling, F. (1992). Personality structure across cultures: A multi method evaluation. *Journal of Personality and Social Psychology.*

Pelcovitz, D., Kaplan, S., Goldenberg, B., & Mandel, F. (1994). Post-traumatic stress disorder in physically abused adolescents. *Journal of American Academy of Child and Adolescents Psychiatry,* 33(3), 305-312.

Pelton, L. (1978). Child abuse and neglect: The myth of class lessens, *American Journal of Orthopsychiatry,* 48(4), 608-617.

Piaget, J. (1972). Intellectual evolution from adolescence to adulthood. *Human Development.* Vol.15, 1-12.

Pierce, R., & Pierce, L.H. (1985). The sexually abused child: A comparison of male and female victims. *Child Abuse & Neglect,* 9, 191-199.

Piers, E.V. (1973). Adolescent creativity. In J.F. Adams (Ed.), *Understanding Adolescence: Current Developments in Adolescent Psychology,* 2d Ed. Bosten: Allyn & Bacon, pp. 191-220.

Pelton, L.H. (1994). The role of material factor in child abuse and neglect. In G.B. Melton & F.D. Barry (Eds), *Protecting Children from Abuse and Neglect* (pp. 166-181). New York: Guilford.

Prino, C.T. & Peyrot, M. (1994). The effect of child physical abuse and neglect on aggressive, withdrawn, and pro-social behavior. *Child Abuse & Neglect*, 18, 871-884.

Pulski, M.A.S. (1970). Play as a function of toy structure and fantasy predisposition. *Child Development*, 41, 531- 537.

Raynor, J.O. (1970). Relationships between achievement-related motives, future orientation, and academic performance. *Journal of Personality and Social Psychology*, 15, 28-33.

Read, J.D. & Lindsay, D.S. (1997). *Recollections of Trauma: Scientific Evidence and Clinical Practice*. New York: Plenum.

Reider, C. & Cicchetti, D. (1989). Organizational perspective on cognitive control functioning and cognitive affective balance in maltreated children. *Developmental Psychology*, 25, 382-393.

Reider, P., Duncon, S. & Gray, M. (1993). Beyond Blame: Child Abuse Tragedies Revisited. London: Routledge.

Rogoff, B. (1990). *Apprenticeship in Thinking*. New York: Oxford University Press.

Rohner, R.P. (1980). Word wide test of parental acceptance-rejection theory. *An Overview Behavior Science Research*, 15, 1-12.

Rohner, R.P. & Rohner, E.C. (1980). Antecedents and consequences of parental rejection: A theory of emotional abuse. *Child Abuse & Neglect*, 4, 189-198.

Rosen, B.C. (1962). Race, socialization and the achievement motivation in Brazil. *American Sociological Review*, 27, 612-624.

Rushton, J.P. & Teachman, G. (1978). The effects of positive reinforcement. Attributions, and punishment on model induced altruism in children. *Personality and Social Psychology Bulletin*, 4, 322-325.

Russel, D.E.H. (1986). *The Secret Trauma: Incest in the Lives of Girls and Woman*. New York: Basic Books.

Rust. J.D. & Troupe, P.A. (1991). Relationship of treatment of child sexual abuse with school achievement and self-concept. *Journal of Early Adolescence*, 11, 420-429.

Salzinger, S., Kaplan, S., Pelcoviz, D., Samit, C. & Krieger, R. (1984). Parent and teacher assessment of children's behavior in child maltreating families. *Journal of the American Academy of Child Psychiatry*.

Santrock, J.W. & Wohlford, P. (1970). Effects of father absence: Influence of the reason for and the onset of absence. *Proceedings of the Annual Convention of American Psychological Association,* 5, 265-266.

Scheintuch, G. & Lewin, G. (1981). Parents attitudes and children deprivation: child-rearing attitudes of parents as a key to the advantaged-disadvantaged distinction in pre-school children. *International Journal of Behavioral Development.* 4, 125-142.

Schteingart, J. S. Molnar, J., Klein, T.P., Lowe, C.B. & Hartman, A. H. (1995). Home less ness and child functioning in the context of risk and protective factors moderating child outcomes. *Journal of Clinical Child Psychology*, 24, 320-331.

Sedlak, A.J. & Broadhurst, D.D. (1996). *Executive Summary of the Third National Incidence Study of Child Abuse and Neglect.* (DHHS Pub. No. ACF 105-94-1840). Washington, D.C.; Government Parenting Office.

Selye, H. (1956). *The Stress of Life*. New York: McGraw-Hill.

Sharma, S., Spielberger, H.D. & Singh, M. (1973). Development of the Hindi Edition of the State Trait Anxiety Inventory. *Indian Journal of Psychology*, Vol. 48.

Shepard, R.N. (1957). Stimulus and response generation: A stochastic model relating generation to distance in psychological space. *Psychometrika*, 22, 325-345.

Shields, A. M., Cicchetti, D. & Ryan, R. M. (1994). The development of emotional and behavioral self-regulation and social competence among maltreated school age children. *Development and Psychopathology*, 6, 57-75.

Sinha, D. (1976). *Social Disadvantages and Development of Certain Perceptual Skill.* Paper presented at IPA International Seminar on Psychology of Cognition held at Delhi University, March.

Sinha, D. (1984). *Manual for Story Pictorial E.F.T. and Indo-African E.E.T.* Rupa Psychological Center. Bhelupur, Varanasi.

Slade, B.B., Steward, M.S., Morrison, T.L., & Abramowitz. S.I. (1984). Locus of control. Persistence and use of contingency information in physically abused children. *Child Abuse & Neglect*, 8, 447-457.

Spinetta, J.J. & Rigler, D. (1972). The child-abusing parent: A psychological review. *Psychological Bulletin*, 77, 296-304.

Srivastava, P.K. & Tiwari, M.L. (1967). Socio-economic stratification of need achievement. *Psychological Studies*, 12, 9-16.

Srivastava, S. (2002). Work stress, coping and health status among career oriented females. *Unpublished Doctoral Dissertation.* D.D.U. Gorakhpur University, Gorakhpur.

Steele, B. (1980). *Psychodynamic Factors in Child Abuse.* In C.H. Kempe and R.E. Helfer (Eds). The battered child, 3rd ed., Chicago: University of Chicago Press.

Stokols, D. (1972). On the distinction between density and crowding. Some implications for further research. *Psychological Review,* 79, 275- 277.

Strakey, S.L. (1980). The relationship between parental acceptance-rejection and the academic performance of fourth and fifth graders. *Behavior Science Research,* 15, 67-80.

Straus, M. (1980). Stress and physical abuse. *Child Abuse and Neglect,* 4, 75-88.

Straus, M.A. & Kantor, G.K. (1994). Corporal punishment of adolescents by parents: A risk factor in the epidemiology of depression, suicide, alcohol abuse child abuse and wife beating. *Adolescence,* 29, 543-561.

Straus, M.A. (1977). Societal morphogenesis and intrafamily violence in cross-cultural perspective. *Annals of the New York, Academy of Sciences,* 285, 717-730.

Straus, M.A., & Gelles, R.J. (Eds), (1990). *Physical Violence in American Families.* New Brunswick. N.J.: Transaction Books.

Straus, M.A., Gelles, R.J. & Steinmetz, S.K. (1980). *Behind Closed Doors.* Beverly Hills, C.A.: Sage.

Suri, I.S. (1989). An investigation in to the structure of reasoning ability of 18-year-old students belonging to rural and urban areas, Ph.D., Edu. Jamia Millia Islamia.

Taylor, S. E. (1991). *Health Psychology* (2nd ed) New York. McGraw-Hill.

Thibaunt, J.W. & Kelley, H.H. (1959). *The Social Psychology of Groups.* New York: Wiley.

Thompson, R.A. (1994). Social support and the prevention of child maltreatment. In G.B. Metton & Barry (Eds). *Protecting Children from Abuse and Neglect: Foundations for New National Strategy* (pp. 40-130). New York: Guilford.

Times of India, (2004, November).

Tolan, P.H. & Guerra, N. (1998). Societal causes of violence against children. In P.K. Trickett & C.J. Schellenbach (Eds.), *Violence Against Children in the Family and the Community,* (pp.195-209). Washington D.C.: American Psychological Association.

Torrance, E.P. (1972). Creative young women in today's world. *Exceptional Children*, 38, 597-603.

Travis, C. (1993, January). Beware the incest-survivor machine. *New York Times*, p.1.

Triandis, H.C. (1995). *Individualism and Collectivism*. Boulder, Co: Westview Press.

Trickett, P.K., Aber, J.L., Carlson, V. & Cicchetti, D. (1991). Relationship of socio economic status to the etiology and developmental sequelae of physical child abuse. *Developmental Psychology*, 27, 148- 158.

Trickett, P.K., & Susman, E.J. (1988). Parental perceptions of child-raising practices in physically abusive & non-abusive families. *Developmental Psychology*. 24, 270-276.

Tripathi, L.B. (1988). Higher mental process. In J. Pandey (Ed). *Psychology in India: The State of the Art*. Vol-1, New Delhi: Sage.

Trivers, R.L. (1971). The evolution of reciprocal altruism. *Quarterly Review of Biology*, 46, 35-57.

Trocime, N., McPhee, D., Tam, K.K., & Hay, T. (1994). *Ontario Incidence Study of Reported Child Abuse and Neglect*. Toronto: Institute for the Prevention of Child Abuse. (Available from the Institute for the Prevention of Child Abuse, 25 Spadina Ave, Toronto, ont. M 5R, 259.)

Trommsdorf, G.C. (1983). Future orientation and socialization. *International Journal of Psychology*, 18, 381-406.

Tulving, E. (1962). Subjective organization in free recall of "unrelated word." *Psychological Review*, 69, 344-354.

Tulving, E. (1966). Subjective organization and effects of repetition in multi trial free recall verbal learning. *Journal of Verbal Learning and Verbal Behavior*, 5, 193-197.

Uhr, R., Thamas, H. & Becker, J. (1969). Entwich-lungsverlanfe in kindes und Jugenalter. Zeischrift fur entwicklungs psychologee und padagogische psychologie.

Verma, K. (1988). *An Experimental Study of Short-Term and Long-Term Memory as Determined by Motivation, Age, Sex and Sense Modalities*, Ph.D., Psy. Panjab University.

Verma, M., Upadhyay, S.K. & Mishra, R.C. (2003). Effect of task structure on the development of perceptual exploration. *Social Science International,* Volume 19.

Vissing, Y. M., Straus, M. A., Gelles, R. J. & Harrop, J. W. (1991).

Verbal aggression by parents and psychosocial problems of children. *Child Abuse & Neglect*, 15, 223-238.

Vondra, J.A., Barnett, D. & Cicchetti, D. (1990). Self-concept, motivation and competence among preschoolers from maltreating and comparison families. *Child Abuse & Neglect*, 14, 525-540.

Weiner, B. (1982). The emotional consequences of causal attributions. In M.S., Clark & S.T., Fiske (Eds). *Affect and Cognition: The 17th Annual Carnegie Symposium on Cognition*, (pp. 185-210). Hillsdale, NJ: Erlbaum.

Whiteman, M., & Deutsch, M. (1968). Social disadvantage as related to intellective and language development. In Deutsch, M. *et al.* (Eds). *Social Class, Race and Psychological Development*. N.Y.: Holt, pp. 86-114.

Whitmore, E.A.W., Kramer, J.R. & Knutson, J.F. (1993). The association between punitive childhood experiences and hyperactivity. *Child Abuse & Neglect,* 17, 357-366.

Widom, C.S. (1989). Dose violence beget violence? A critical examination of the literature. *Psychological Bulletin*, 106, 3-28.

Wilkinson, J.E., & Murphy, H.F. (1976). Differential method of enhancing growth in urban preschool children. *Child Care & Health Development*, 1-11.

Williams, L.M. & Finkelhor, D. (1990). The characteristics of incestuous fathers: A review of recent studies. In W. L. Marshall, D.R. Lawss & H.E. Barbaree (Eds). *Handbook of Sexual Assault: Issue, Theories and Treatment of the Offender* (pp. 231-255). New York: Plenum.

Wind, T. W. & Silvern, L. (1992). Type and extent of child abuse as predictors of adult functioning. *Journal of Family Violence*, 7, 261-281.

Witkin, H.A. & Goodenough, D.R. (1981). *Cognitive Styles: Essence and Origins.* New York: International University Press.

Witkin, H.A., Dyk, R.B., Faterson, H.F., Goodenough, D.R. & Karp, S.A. (1962). *Psychological Differentiation*, N.Y.: Wiley, 1962.

Wodarski, J. S., Kurtz, P. D., Gaudin, J. M., & Howing, P.T. (1990). Maltreatment and the school age child: Major academic, socio-emotional and adaptive outcomes. *Social Work*, 35, 506-513.

Wolfe, D. (1987). *Child Abuse: Implications for Child Development and Psychopathology,* Newbury Park. CA. Sage.

Wolfe, D. (1999). *Child Abuse: Implications for Child Development and Psychopathology* (2nd ed.) Thousand Oake, CA: Sage.

Wolfe, D. & Manion, I. (1984). Impediments to child abuse prevention:

Issues and directions. *Advances in Behavior Research and Therapy*, 6, 47- 67.

Wolfe, D., Werkele, C., Reitzcl-Jaffe, D. & Lefebvere, L. (1998). Factors associated with abusive relationship among maltreated and non-maltreated youth. *Development and Psychopathology*, 10, 61-85.

Wolfe, D.A., Jaffe, P. J., Wilson, S. K. & Zak, L. (1985). Children of battered women: The relation of child behavior to family violence and maternal stress. *Journal of Consulting and Clinical Psychology*, 53, 657-665.

Wolfe, D. A. (1985). Child abusive parents. An empirical review and analysis. *Psychological Bulletin*, 97, 462-482.

Wolfgang, M.E. & Ferracuti, F. (1972). *Delinquency in a Birth Cohort*. London: Tavistock.

World Health Organization (1997, March). *Child Abuse and Neglect. http://www.who.int/inf-fs/en/fact. 150.html*

Wyatt, G.E. (1994). Socio-cultural and epidemiological issues in the assessment of domestic violence. *Journal of Social Distress and the Homeless*, 3, 7-21.

Zahn–Waxler, C., Cole, P.M., Welsh, J.D. & Fox, N.A. (1995). Psychological correlates of empathy and pro-social behavior in preschool children with behavior problems. *Development and Psychopathology*, 7, 27-48.

Zigler, E. (1980). Controlling child abuse: Do we have the knowledge and/or the will? In G. Gerbner, C. Ross & E. Zigler (Eds.). *Child Abuse : An Agenda for Action*. New York : Oxford.

Zillman, D. (1979). Hostility and aggression. Hillsdale, N.J. Lawrence Erlbaum.

Zimbardo, P.G. (1970). The human choice: Individuation reason and order versus deindividuation, impulse and chaos. In N.J. Arnold & Levine (Eds.). *Nebraska Symposium on Motivation*, 1969. Lincoln: University of Nebraska Press.

Author Index

Subject Index